Superalloys

A Technical Guide

Elihu F. Bradley

Consulting Editor

ASM INTERNATIONAL™
Metals Park, OH 44073

First printing, May 1988
Second printing, November 1989

Library of Congress Catalog Card Number: 88-070147 ISBN: 0-87170-327

Printed in the United States of America

ASM INTERNATIONAL
Metals Park, OH 44073
Tel: (216) 338-5151

Preface

The advent of the aircraft turbine engine in the late 1940's and 1950's gave great impetus to high-temperature materials. The concept of the jet engine dictated the need for new construction materials that would combine high-temperature, long-time strength with resistance to elevated temperature corrosion. Throughout the age of the aircraft turbine engine, there has been continuing demand for new materials with improved high-temperature performance, because of the simple fact that the higher the turbine inlet temperature, the higher the thrust that the engine can produce. This demand has fueled a burgeoning superalloy industry throughout the years. It is revealing that the amount of superalloys used in aircraft gas turbine engines has increased from about 10% of the total engine weight at the beginning of the jet age in 1950 to about 50% of the engine weight for modern engines. It is expected that this trend will continue in the future, probably leveling off at about 60% in the 1990's. The use of superalloys has spread beyond aircraft engines as other industries have recognized their high-temperature advantages, but the turbine engine remains the cornerstone of superalloy use.

The physical metallurgy of superalloys is well understood and well documented in the technical literature. It is not the purpose of this book to present a metallurgical treatise on superalloys, because there are many in the current literature. Rather, the intent is to provide the technical community, but not necessarily superalloy specialists, with a current, practical overview or summary of what superalloys are all about. Thus, superalloy types, forms, properties, applicable processes, uses, and problems are covered. The emphasis, therefore, is on practical, not theoretical, information concerning superalloys. Technology, however, is included to support the practical information. Every attempt has been made to express the technical sections in clear, easily understood terms, thus avoiding complex discussions of chemistry and solid-state physics.

It is hoped, then, that this book will be used by producers, fabricators, manufacturers, designers, metallurgists, laboratory personnel, and users of superalloys as a broad, practical reference to superalloy information required for application purposes.

E.F. Bradley

West Hartford, CT
May 1988

Contents

Chapter 7

Metallography 99

Chapter 8

Castings 109

Chapter 9

Forging 133

Chapter 12

Cleaning, Finishing 185

Chapter 13

Welding 197

Chapter 14

Brazing 221

Chapter 15

Protective Coatings. 233

Chapter 16

Identification, Recycling. 241

Chapter 17

Failure Analysis and Prevention 249

Chapter 1

Introduction

GENERAL

Many alloys are used at elevated temperatures. These alloys must be able to withstand the deteriorating effect of the service atmosphere, as well as possess sufficient strength for the design condition and have adequate stability to withstand damaging metallurgical structural changes at operating temperature. From the standpoint of resisting oxidation and high-temperature corrosion, the most important alloying element is chromium. It is not surprising that corrosion-resistant steels, stainless steels, nickel-chromium alloys, and superalloys, all of which contain significant amounts of chromium, are used extensively in high-temperature applications.

This book is intended to cover only superalloys, because stainless steels are discussed extensively elsewhere (Ref 1). However, the following brief comments about materials that are used at elevated temperatures other than superalloys are given to help the reader better understand the application role of superalloys. For use at moderate temperatures – less than 540 °C (1000 °F) – and moderate stress, the 12% Cr corrosion-resistant steels are satisfactory. Under conditions of somewhat higher stress at the same moderate temperature, the so-called super 12% Cr steels, a versatile group containing in addition to chromium small amounts of molybdenum and/or other strong carbide formers and/or cobalt or nickel, have been used for over 40 years. As the operating temperature is increased, but for conditions of low stress, higher chromium steels, either the ferritic corrosion-resistant or the austenitic stainless steels containing nickel as well as chromium, or nickel-chromium alloys (for example, Nichrome V, containing 20% Cr and 80% Ni), commonly are selected.

For very high operating temperatures, there is increasing interest in the refractory metals of Groups V (vanadium, niobium, and tantalum) and VI (chromium, molybdenum, and tungsten), as well as ceramics. The refractory metals, however, exhibit very poor oxidation resistance, and their use is now restricted to nonoxidizing environments. Ceramics possess insufficient toughness for most structural applications. Elevated temperatures have been a disappointing limitation for titanium ever since its introduction into turbine engines in the middle 1950's. Two reasons for this limitation persist: (*a*) the affinity of titanium for interstitial elements, and (*b*)

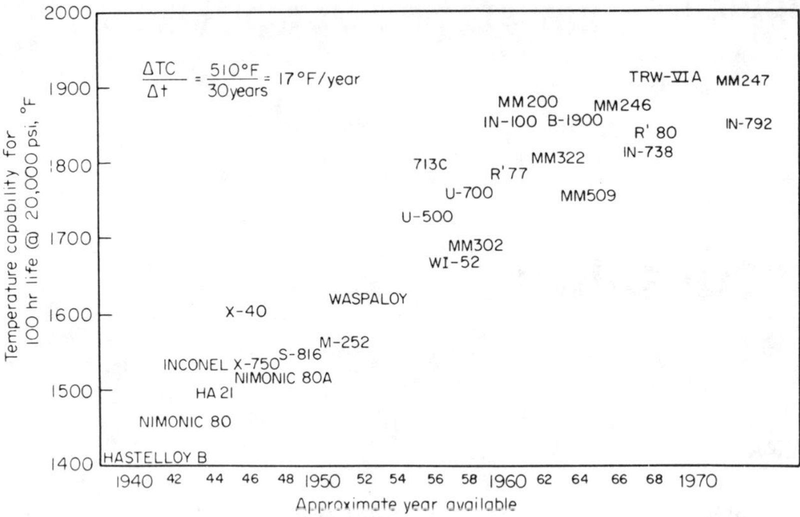

Fig. 1.1 Historical development and typical temperature capability of superalloys

inadequate creep strength at even the moderate temperature level of about 540 °C (1000 °F). Therefore, for the most severe combination of stress and temperature, it remains for the remarkable superalloys to do the job.

WHAT IS A SUPERALLOY?

A superalloy is an alloy developed for elevated temperature service, usually based on Group VIII A elements, where relatively severe mechanical stressing is encountered and where surface stability frequently is required. The term "superalloy" was first used shortly after World War II to describe a group of alloys developed for use in turbosuperchargers and aircraft turbine engines that required high performance at elevated temperatures. These alloys usually consist of various formulations made from the following elements: iron, nickel, cobalt, and chromium, as well as lesser amounts of tungsten, molybdenum, tantalum, niobium, titanium, and aluminum. The most important properties of the superalloys are long-time strength at temperatures above 650 °C (1200 °F) and resistance to hot corrosion and erosion. Many types of alloys fall under the broad coverage of superalloys. These include iron-base alloys containing chromium and nickel, complex iron-nickel-chromium-cobalt compositions, carbide-strengthened cobalt-base alloys, solid-solution-strengthened nickel-base alloys, and precipitation- or dispersion-strengthened nickel-base alloys (Fig. 1.1). Superalloys are used in both the wrought and the cast forms.

Generally, the strengths of the iron-base alloys, the complex iron-nickel-chromium-cobalt alloys, and the nickel-base solid-solution-strengthened alloys are considerably lower than those of the nickel-base second-phase-strengthened and cobalt-base alloys at temperatures above 650 °C (1200 °F). Early iron-base superalloys,

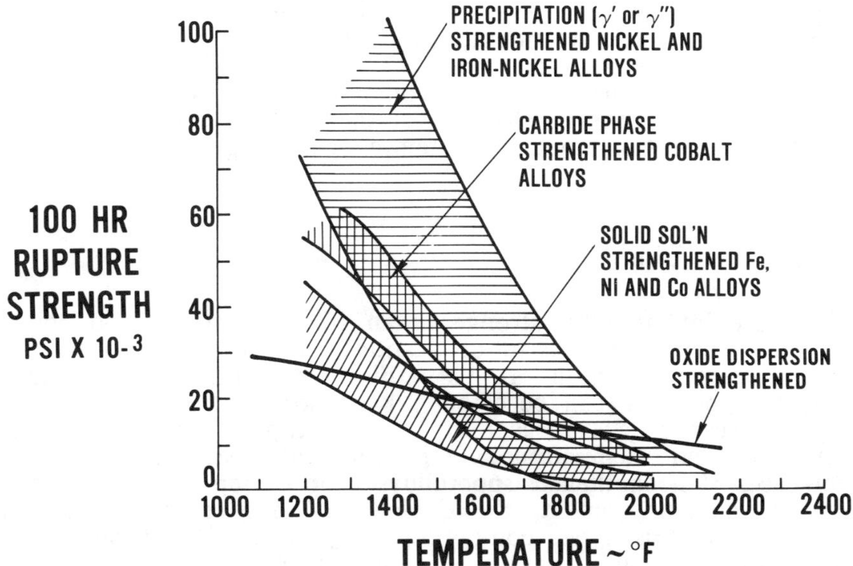

Fig. 1.2 Stress-rupture characteristics of select superalloys

such as 16-25-6 alloy containing 16% Cr, 25% Ni, and 6% Mo, and complex iron-nickel-chromium-cobalt alloys (Fe-20Ni-20Cr-20Co, for example) with small amounts of tungsten and molybdenum, are essentially solid-solution strengthened. Later iron-base alloys, containing small amounts (2 to 3%) of aluminum and titanium, achieved increased high-temperature strength through precipitation of an aluminum-titanium strengthening phase. Because of a melting-point advantage, the cobalt alloys are usually stronger than the nickel alloys at temperatures above 1100 °C (2000 °F). Cast cobalt-base alloys, characterized by a face-centered-cubic (fcc, austenitic) solid-solution matrix and containing complex carbides, have had a successful history as airfoils for gas turbine engines (most turbine vanes and some turbine blades).

One exception to this strength observation is the dispersion-strengthened nickel alloys, utilizing a dispersed-oxide strengthening phase, which exhibit high strength at elevated temperatures, but only moderate strength at intermediate temperatures. The secondary phase persists in the structure of these alloys as a strengthening mechanism throughout the solid state until melting occurs. In contrast, alloys strengthened by precipitation lose the strengthening phase by solution in the solid state at some temperature below the melting point. Dispersion-strengthened alloys are beginning to be used in some gas turbine engine burner applications. Figure 1.2 compares stress-rupture behavior of the various superalloy classes.

Clearly, the nickel-base superalloys strengthened by a secondary precipitated phase are the most complex and, indeed, the most remarkable of all the superalloys. The physical metallurgy of these alloys is subtle, sophisticated, and well understood. The structure consists of an fcc, austenitic solid-solution matrix with a precipitated nickel-aluminum-titanium compound (γ') as the principal strengthening phase. Various carbides, depending on the particular alloy composition and heat treatment, exist as second precipitated phases. These alloys are used in the most demanding applications relative to stress and temperature in gas turbine engines. They have demonstrated remarkably useful strength at the highest fraction of the base metal melting point of any alloy system ever developed.

Castings and Forgings

Castings are intrinsically stronger than forgings at elevated temperatures. The coarse grain size of castings, as compared to fine-grain forgings, favors strength at very high temperatures. In addition, casting compositions can be effectively tailored for high-temperature strength, inasmuch as forgeability characteristics are not a factor. Higher elevated temperature strength can be achieved in the nickel-base, γ'-strengthened superalloys, for example, by lowering the chromium content, but at the expense of hot corrosion resistance. Superalloys of this type, containing only 8 to 12% chromium, may require the use of coatings (such as diffused aluminum) to compensate for the loss in hot corrosion resistance of the alloy. The addition of small amounts of hafnium (1 to 1.5%) causes a marked improvement in the intermediate-temperature ductility of these high-strength nickel superalloy castings. Because of the high-temperature strength advantage, many aircraft gas turbine engines use nickel-base, γ'-strengthened superalloy castings for high-stress, high-temperature turbine blade applications.

Innovation in the casting process also has resulted in improved high-temperature properties. The development of directional solidification involving controlled grain growth, whereby all crystals are aligned in the longitudinal direction, has provided increased high-temperature strength and, in particular, significant improvement in resistance to thermal fatigue.

Wrought Products

The major structural difference between a wrought product and a casting is that wrought products possess a finer grain condition, which is achieved by hot working. Wrought products have better strength and ductility than castings at room-to-moderately high temperatures – about 540 °C (1000 °F). Wrought products, in addition, generally have better fracture and fatigue properties than castings, because defects and large grains are broken up and porosity is healed during the hot working processes. Therefore, for critical structural applications requiring dynamic fracture reliability at these low-to-moderate temperatures, the material choice has favored wrought products.

Powder Metallurgy

The use of metal powders to obtain chemically and metallurgically uniform structures for critical aircraft engine parts is increasing with the advent of high-purity, prealloyed superalloy powders, and the development of suitable shaping techniques, such as isothermal forging and hot isostatic pressing. Another significant development relative to powder cleanliness is inert processing, wherein powder production, collection, and densification are carried out in an inert atmosphere. High-strength, nickel-base superalloys are prone to severe macrosegregation that inhibits successful ingot breakdown. Conceptually, powder metallurgy (P/M) offers a method for overcoming this problem. Because the material is divided into small droplets while it is a homogeneous liquid, the maximum segregation distance is restricted to the size of the solidified droplets.

Production of superalloy powders using rapid solidification technology (RST) is receiving major research attention. In view of the high cooling rate achieved in RST - 1,008,000 °C/s (1,800,000 °F/s) - it may be possible to produce new alloys and microstructures with a high degree of compositional homogeneity and fine microstructure in each spherical powder particle with very narrow and fine size distribution. This could provide a quantum leap in alloy technological potential. The value of these metastable microstructures, however, will be determined by the extent to which the metastability persists during processing and subsequent service.

NEW APPROACHES

Although materials technology has pushed the use of superalloys to temperatures close to their melting points, materials scientists are studying methods for still further advances. Single-crystal components, characterized by the complete absence of grain boundaries, are beginning to be used. Directionally solidified eutectics, wherein aligned whiskers grow from the eutectic phase within a ductile matrix, are emerging as fiber-reinforced alloys (*in situ* composites) of great strength and stability with the promise of even higher temperature capability. New approaches involving P/M and thermomechanical processing may significantly improve strength at intermediate temperatures. The use of P/M technology to disperse oxide particles in γ'-strengthened superalloys has promise of improved high-temperature strength. The combination of composite technology with superalloy metallurgy may provide future turbine blade materials.

APPLICATIONS

The high-temperature applications of superalloys are extensive, including components for aircraft, chemical plant equipment, and petrochemical equipment. The increasing significance of superalloys in today's commerce is typified by the fact that, whereas in 1950 only about 10% of the total weight of an aircraft turbojet engine was made of superalloys, by 1985 this figure had risen to about 50%, and it is expected to reach 60% by 1993.

Current applications of superalloys include:

Aircraft and industrial gas turbines:
- Disks
- Bolts
- Shafts
- Cases
- Blades
- Vanes
- Burner cans
- Afterburners
- Thrust reversers

Steam-turbine power plants:
- Bolts
- Blades
- Stack-gas reheaters

Reciprocating engines:
- Turbochargers
- Exhaust valves
- Hot plugs
- Precombustion cups (Riccardo-type diesel)
- Valve-seat inserts

Metal processing:
- Hot work tools and dies
- Cast dies

Medical applications:
- Dentistry
- Prosthetic devices

Space vehicles:
- Aerodynamically heated skins
- Rocket-engine parts

Heat treating equipment:
- Trays
- Fixtures
- Conveyor belts

Nuclear power systems:
- Control-rod drive mechanisms
- Valve stems
- Springs
- Ducting

Chemical and petrochemical industries:
- Bolts
- Valves
- Reaction vessels
- Piping
- Pumps

PRODUCT FORMS

Wrought superalloys are manufactured in all mill forms common to the metal industry. Iron-base, cobalt-base, and nickel-base superalloys are produced conventionally as bar, billet, extrusions, plate, sheet, strip, wire, and forgings by primary mills. Superalloys also are available as rod, bar, plate, sheet, strip, tube, pipe, shapes, wire, forging stock, and specialty items from secondary converters.

High-purity, prealloyed superalloy powders are available from a number of sources as the starting stock for isothermal forging, superplastic shaping, or hot isostatic pressing of components. Investment cast superalloys are used widely in aerospace applications for turbine airfoils and in large structural parts.

REFERENCE

1. *Stainless Steel,* revised by R.A. Lula, American Society for Metals, 1986.

Chapter 2

Metallurgy of Superalloys

INTRODUCTION

It is not the purpose of this chapter to present an in-depth metallurgical treatise on superalloys; there are many available in the current literature (Ref 1, 2, and 3). Superalloy technology, however, is so important to the foundation of this book that a basic technological overview is included.

Superalloys are intended for elevated temperature service at temperatures of 650 °C (1200 °F) and above where relatively severe mechanical stressing is encountered and where surface stability often is required. The term "superalloy" can be applied to a wide variety of iron-, nickel-, and cobalt-base materials, or complex combinations of iron, nickel, cobalt, and chromium. These alloys consist of various formulations of iron, nickel, cobalt, and chromium, as well as lesser amounts of tungsten, molybdenum, tantalum, niobium (also called columbium), titanium, and aluminum. Nickel superalloys may also contain small amounts of boron, zirconium, and hafnium, which enhance creep-rupture life and ductility through grain-boundary morphology.

Carbon is present in all superalloys, usually up to about 0.03% in nickel and iron superalloys, but may be present in higher amounts in cobalt-base alloys to induce carbide-phase strengthening. The various types of superalloys are iron base with chromium and nickel, complex iron-nickel-chromium-cobalt compositions, solid-solution-strengthened cobalt-base alloys, cobalt-base carbide-strengthened alloys, nickel-base solid-solution-strengthened alloys, nickel-base precipitation-strengthened alloys, and nickel-base oxide-dispersion-strengthened alloys. Examples of these alloys are listed in Tables 2.1 and 2.2.

Superalloys are used in both the wrought and cast forms. Wrought product forms may be made by conventional mill processes starting from the ingot, or by powder metallurgy (P/M) processing. Castings generally are made by the precision investment process.

Table 2.1 Nominal compositions of wrought superalloys

Alloy	UNS No.	Composition, %										
		Cr	Ni	Co	Mo	W	Nb	Ti	Al	Fe	C	Other
Iron-base solid-solution alloys												
16-25-6	...	16.0	25.0	...	6.0	...	...	...	...	50.7	0.06	1.35 Mn, 0.70 Si, 0.15 N
Carpenter 20Cb-3	N08020	20.0	34.0	...	2.50	...	1.0 max	...	...	42.4	0.07 max	3.5 Cu
Incoloy 800	N08800	21.0	32.5	...	...	...	...	0.38	0.38	45.7	0.05	...
Incoloy 801	N08801	20.5	32.0	...	...	...	...	1.13	...	46.3	0.05	...
Incoloy 802	...	21.0	32.5	...	...	...	...	0.75	0.58	44.8	0.35	...
N-155	R30155	21.0	20.0	20.0	3.00	2.5	1.0	...	...	32.2	0.15	0.15 N, 0.02 La, 0.02 Zr
RA-330	N08330	19.0	36.0	...	...	...	...	...	...	45.1	0.05	...
Iron-base precipitation-strengthening alloys												
A-286	K66286	15.0	26.0	...	1.25	...	...	2.0	0.2	55.2	0.04	0.005 B, 0.3 V
Discaloy	K66220	14.0	26.0	...	3.0	...	...	1.7	0.25	55.0	0.06	...
Haynes 556	...	22.0	21.0	20.0	3.0	2.5	0.1	...	0.3	29.0	0.10	0.50 Ta, 0.02 La, 0.002 Zr
Incoloy 903	...	0.1 max	38.0	15.0	0.1	...	3.0	1.4	0.7	41.0	0.04	...
Pyromet CTX-1	...	0.1 max	37.7	16.0	0.1	...	3.0	1.7	1.0	39.0	0.03	...
V-57	...	14.8	27.0	...	1.25	...	...	3.0	0.25	48.6	0.08 max	0.01 B, 0.5 V max
W-545	K66545	13.5	26.0	...	1.5	...	...	2.85	0.2	55.8	0.08	0.05 B
Cobalt-base solid-solution alloys												
Haynes 25 (L-605)	R30605	20.0	10.0	50.0	...	15.0	...	...	...	3.0	0.10	1.5 Mn
Haynes 188	R30188	22.0	22.0	37.0	...	14.5	...	...	...	3.0 max	0.10	0.90 La
J-1570	...	20.0	28.0	46.0	...	...	...	4.0	...	2.0	0.2	...
MAR-M 302	...	21.5	...	58.0	...	10.0	...	...	...	0.5	0.85	9.0 Ta, 0.005 B, 0.2 Zr
MAR-M 509	...	23.5	10.0	54.5	...	7.0	...	0.2	...	...	0.6	0.5 Zr, 3.5 Ta
MP-35N	R30035	20.0	35.0	35.0	10.0	...	...	...	...	...	...	...

(continued)

Table 2.1 (continued)

Alloy	UNS No.	Composition, %										
		Cr	Ni	Co	Mo	W	Nb	Ti	Al	Fe	C	Other
Cobalt-base solid-solution alloys												
MP-159	...	19.0	25.0	36.0	7.0	...	0.6	3.0	0.2	9.0	...	...
S-816	R30816	20.0	20.0	42.0	4.0	4.0	4.0	...	...	4.0	0.38	...
Stellite 6B	...	30.0	1.0	61.5	...	4.5	...	...	...	1.0	1.0	...
UMCo-50	...	28.0	...	49.0	...	...	...	...	...	21.0	0.12 max	...
WI-52	...	21.0	...	63.5	...	11.0	...	...	...	2.0	0.45	2.0 Nb + Ta
X-40	...	22.0	10.0	57.5	...	7.5	...	...	...	1.5	0.50	0.5 Mn, 0.5 Si
Nickel-base solid-solution alloys												
Hastelloy B	N10001	1.0 max	63.0	2.5 max	28.0	...	...	...	...	5.0	0.05 max	0.03 V
Hastelloy B-2	N10665	1.0 max	69.0	1.0 max	28.0	...	...	...	...	2.0 max	0.02 max	...
Hastelloy C	N10002	16.5	56.0	...	17.0	4.5	...	...	...	6.0	0.15 max	...
Hastelloy C-4	N06455	16.0	63.0	2.0 max	15.5	...	...	0.7 max	...	3.0 max	0.015 max	...
Hastelloy C-276	N10276	15.5	59.0	...	16.0	3.7	...	...	...	5.0	0.02 max	...
Hastelloy N	N10003	7.0	72.0	...	16.0	...	...	0.5 max	...	5.0 max	0.06	...
Hastelloy S	...	15.5	67.0	...	15.5	...	...	...	0.2	1.0	0.02 max	0.02 La
Hastelloy W	N10004	5.0	61.0	2.5 max	24.5	...	...	...	...	5.5	0.12 max	0.6 V
Hastelloy X	N06002	22.0	49.0	1.5 max	9.0	0.6	...	...	2.0	15.8	0.15	...
Inconel 600	N06600	15.5	76.0	...	...	...	...	...	...	8.0	0.08	0.25 max Cu
Inconel 601	N06601	23.0	60.5	...	...	...	...	...	1.35	14.1	0.05	0.05 max Cu
Inconel 604	...	16.0	74.0	...	...	...	2.25	...	...	7.5	0.02	0.03 max Cu
Inconel 617	...	22.0	55.0	12.5	9.0	...	...	...	1.0	...	0.07	...
Inconel 625	N06625	21.5	61.0	...	9.0	...	3.6	0.2	0.2	2.5	0.05	...
NA-224	...	27.0	48.0	...	...	6.0	...	...	...	18.5	0.50	...
Nimonic 75	...	19.5	75.0	...	...	...	...	0.4	0.15	2.5	0.12	0.25 max Cu
RA-333	N06333	25.0	45.0	3.0	3.0	3.0	...	...	...	18.0	0.05	...

(continued)

Table 2.1 (continued)

Alloy	UNS No.	Composition, %										
		Cr	Ni	Co	Mo	W	Nb	Ti	Al	Fe	C	Other
Nickel-base precipitation-strengthening alloys												
Astroloy	...	15.0	56.5	15.0	5.25	...	...	3.5	4.4	<0.3	0.06	0.03 B, 0.06 Zr
B-1900	...	8.0	63.3	10.0	6.0	...	...	1.0	6.0	...	...	4.3 Ta, 1.3 Hf, 0.01 B, 0.05 Zr
D-979	N09979	15.0	45.0	...	4.0	4.0	...	3.0	1.0	27.0	0.05	0.01 B
IN 100	N13100	10.0	60.0	15.0	3.0	...	...	4.7	5.5	<0.6	0.15	1.0 V, 0.06 Zr, 0.015 B
IN 102	N06102	15.0	67.0	...	2.9	3.0	2.9	0.5	0.5	7.0	0.06	0.005 B, 0.02 Mg, 0.03 Zr
IN MA-754	...	20.0	78.5	...	...	...	...	0.5	0.3	...	...	0.6 Y_2O_3
IN MA-6000E	...	15.0	68.5	...	2.0	4.0	...	2.5	4.5	...	0.05	1.1 Y_2O_3, 2.0 Ta, 0.01 B, 0.15 Zr
INCO 713	...	14.0	72.5	...	4.5	...	2.0	1.0	6.0	...	...	...
Incoloy 901	N09901	12.5	42.5	...	6.0	...	...	2.7	...	36.2	0.10 max	...
Inconel 706	N09706	16.0	41.5	...	...	...	...	1.75	0.2	37.5	0.03	2.9 Nb + Ta, 0.15 max Cu
Inconel 718	N07718	19.0	52.5	...	3.0	...	5.1	0.9	0.5	18.5	0.08 max	0.15 max Cu
Inconel 751	...	15.5	72.5	...	...	...	1.0	2.3	1.2	7.0	0.05	0.25 max Cu
Inconel X750	N07750	15.5	73.0	...	...	...	1.0	2.5	0.7	7.0	0.04	0.25 max Cu
M252	N07252	19.0	56.5	10.0	10.0	...	...	2.6	1.0	<0.75	0.15	0.005 B
MAR-M 004	...	12.0	69.8	...	4.5	...	2.0	1.0	5.9	...	...	4.4 Ta, 1.3 Hf, 0.01 B, 0.05 Zr
MAR-M 200 + Hf	...	9.0	58.4	10.0	...	12.5	1.0	2.0	5.0	...	...	2 Hf, 0.01 B, 0.05 Zr
MAR-M 246	...	9.0	59.2	10.0	2.5	10.0	...	1.5	5.5	...	...	1.5 Ta, 0.01 B, 0.05 Zr
MAR-M 247	...	8.25	59.0	10.0	0.7	10.0	...	1.0	5.5	<0.5	0.15	0.015 B, 0.05 Zr, 1.5 Hf, 3.0 Ta
Nimonic 80A	N07080	19.5	73.0	1.0	...	...	...	2.25	1.4	1.5	0.05	0.10 max Cu
Nimonic 90	N07090	19.5	55.5	18.0	...	...	...	2.4	1.4	1.5	0.06	...
Nimonic 95	...	19.5	53.5	18.0	...	...	...	2.9	2.0	5.0 max	0.15 max	+B, +Zr

(continued)

Table 2.1 (continued)

Alloy	UNS No.	Composition, %										
		Cr	Ni	Co	Mo	W	Nb	Ti	Al	Fe	C	Other
Nickel-base precipitation-strengthening alloys												
Nimonic 100	...	11.0	56.0	20.0	5.0	...	...	1.5	5.0	2.0 max	0.30 max	+B, +Zr
Nimonic 105	...	15.0	54.0	20.0	5.0	...	...	1.2	4.7	...	0.08	0.005 B
Nimonic 115	...	15.0	55.0	15.0	4.0	...	...	4.0	5.0	1.0	0.20	0.04 Zr
Nimonic 263	...	20.0	51.0	20.0	5.9	...	...	2.1	0.45	0.7 max	0.06	...
Pyromet 860	...	13.0	44.0	4.0	6.0	...	...	3.0	1.0	28.9	0.05	0.01 B
Refractory 26	...	18.0	38.0	20.0	3.2	...	...	2.6	0.2	16.0	0.03	0.015 B
Rene' 41	N07041	19.0	55.0	11.0	10.0	...	...	3.1	1.5	<0.3	0.09	0.01 B
Rene' 80	...	14.0	60.0	9.5	4.0	4.0	...	5.0	3.0	...	0.17	0.015 B, 0.03 Zr
Rene' 95	...	14.0	61.0	8.0	3.5	3.5	3.5	2.5	3.5	<0.3	0.16	0.01 B, 0.05 Zr
Rene' 100	...	9.5	61.0	15.0	3.0	...	...	4.2	5.5	1.0 max	0.16	0.015 B, 0.06 Zr, 1.0 V
Udimet 500	N07500	19.0	48.0	19.0	4.0	...	...	3.0	3.0	4.0 max	0.08	0.005 B
Udimet 520	...	19.0	57.0	12.0	6.0	1.0	...	3.0	2.0	...	0.08	0.005 B
Udimet 630	...	17.0	50.0	...	3.0	3.0	6.5	1.0	0.7	18.0	0.04	0.004 B
Udimet 700	...	15.0	53.0	18.5	5.0	...	...	3.4	4.3	<1.0	0.07	0.03 B
Udimet 710	...	18.0	55.0	14.8	3.0	1.5	...	5.0	2.5	...	0.07	0.01 B
Unitemp AF2-1DA	...	12.0	59.0	10.0	3.0	6.0	...	3.0	4.6	<0.5	0.35	1.5 Ta, 0.015 B, 0.1 Zr
Waspaloy	N07001	19.5	57.0	13.5	4.3	...	...	3.0	1.4	2.0 max	0.07	0.006 B, 0.09 Zr

Table 2.2 Oxide-dispersion-strengthened superalloys

Alloy	Ni	Cr	Y_2O_3	Ti	Al	C	Fe	Others
Inconel MA 754	rem	20	0.6	0.5	0.3	0.05	...	
Incoloy MA 956	...	20	0.5	0.5	4.5	...	rem	
Inconel MA 6000E	rem	15	1.1	2.5	4.5	0.05	...	2 Mo, 4 W, 2 Ta, 0.15 Zr, 0.1 B
HDA 8077	rem	16	...	...	4.0	...	...	...
IN 738 + Y_2O_3	rem	16	1.3	3.4	3.4	0.17	...	1.7 Mo, 1.7 Ta, 2.6 W, 0.9 Nb, 8.5 Co, 0.1 Zr

Iron-Base Superalloys

Generally, the high-temperature strengths of the iron-base, complex iron-nickel-chromium-cobalt and the nickel-base solid-solution-strengthened alloys are lower than those of the nickel-base precipitation-strengthened and the cobalt-base carbide-phase-strengthened alloys at temperatures above 650 °C (1200 °F). Early iron-base superalloys – for example, 16-25-6 alloy containing 16% Cr, 25% Ni, 6% Mo, and balance iron, and iron-nickel-chromium-cobalt alloys such as Multimet (or N-155) containing 40% Fe, 20% Ni, 20% Cr, 20% Co, and small amounts of tungsten and molybdenum – were essentially solid solution strengthened. A solid solution, of course, is when two or more metals, or intermetallics, are completely and homogeneously combined with each other in the solid state. A solid solution is the same homogeneous combination as a liquid solution such as sugar in coffee, but in the solid state. Solutions differ from mechanical mixtures, such as a mixture of dry flour and sugar, in which each component retains its individual characteristics. Strengthening occurs when the dissolving of one metal in the other stiffens the new resulting metal.

Later iron-base alloys containing nickel, chromium, and small amounts of aluminum and titanium (1 to 3%) achieved increased high-temperature strength through precipitation of a nickel-aluminum-titanium strengthening phase. The presence of a second phase in the structure is more effective in strengthening than the solid solution. The makeup of the second phase is important – the finer and the more widely distributed, the more effective the strengthening. Indeed, high-temperature exposure causes coalescing of the second phase, which results in larger particles that are less widely distributed, with a resulting loss in strength.

Nickel-Base Superalloys

Nickel-base superalloys are both solid solution and precipitation strengthened. Hastelloy X, an example of the lower strength nickel-base solid-solution superalloys containing 18.5% Fe, 22% Cr, 9% Mo, 1.5% Co, and 0.6% W, has been used widely in burner and combustor applications in gas turbine engines. It exhibits high-temperature corrosion/erosion resistance combined with excellent fabricability and weldability.

For applications requiring high-temperature strength, the precipitation-strengthened nickel-base superalloys usually are selected. Clearly, these alloys, which are strengthened by a secondary precipitated phase, are the most complex and, indeed, the most remarkable of all the superalloys. The physical metallurgy of these alloys is subtle and sophisticated, with the relationship of properties to microstructure well known and understood. These alloys exhibit a face-centered cubic (fcc) austenitic solid-solution matrix or background metal with a precipitated nickel-aluminum-titanium compound (known as γ') as the principal strengthening phase. "Austenitic" is the term given to a solid solution of metals or intermetallics and the base metal of an fcc crystalline structure, i.e., each metal crystal is in the form of a cube with an atom at each corner as well as in the center of each of the six cube faces. Nickel and cobalt are such metals.

Various carbides, depending on the particular alloy composition and heat treatment, also exist as secondary precipitated phases. Without a doubt, the γ'-strengthened nickel superalloys are used in the most demanding applications relative to stress and temperature in gas turbine engines. They have demonstrated remarkably useful strength at the highest fraction of the base-metal melting point of any alloy system ever developed. Another strengthening mechanism in nickel alloys (also used in iron alloys) is the presence of dispersed oxide particles in the structure. Such dispersion-strengthened alloys exhibit the highest strength at the most elevated temperatures, but only low-to-moderate strength at intermediate temperatures.

Because of a melting point advantage, cobalt alloys usually are stronger than the nickel alloys at very high temperatures - above 1095 °C (2000 °F). Cobalt superalloys, like nickel alloys, are characterized by an fcc austenitic solid-solution matrix, and some contain appreciable carbon contents to provide complex carbides as second-phase strengtheners. These cobalt alloys are used extensively in the cast condition for high-temperature turbine vane applications in aircraft gas turbine engines. Other cobalt superalloys, such as L-605 containing 10% Ni, 20% Cr, and 15% Mo, are only solid-solution strengthened. For a review of the elevated temperature strength relationships of the various superalloy classifications, see Fig. 1.2 in Chapter 1.

Castings and Forgings

Castings are intrinsically stronger than forgings at elevated temperature. The coarse grain size of castings, as compared to fine-grain forgings, favors strength at high temperatures. In addition, casting compositions can be tailored effectively for high-temperature strength, inasmuch as forgeability characteristics are not applicable. For example, higher strength at elevated temperatures can be achieved in the nickel-base γ'-strengthened superalloys by increasing aluminum and titanium, thus increasing the volume percentage of γ', and by decreasing the chromium content.

Table 2.3 Effect of elements on superalloys

Effect(a)	Fe-base	Co-base	Ni-base
Solid-solution strengtheners	Cr, Mo	Nb, Cr, Mo, Ni, W, Ta	Co, Cr, Fe, Mo, W, Ta
Face-centered cubic matrix stabilizers	C, Ni, Co	Ni	Co
Carbide form:			
MC type	Ti	Ti	W, Ta, Ti, Mo, Nb
M_7C_3 type	...	Cr	Cr
$M_{23}C_6$ type	Cr	Cr	Cr, Mo, W
M_6C type	Mo	Mo, W	Mo, W
Carbonitrides:			
M(CN) type	C, N	C, N	C, N
Promotes general precipitation of carbides	P	...	...
Forms γ' Ni_3 (Al,Ti)	Al, Ni, Ti	...	Al, Ti
Retards formation of hexagonal η (Ni_3Ti)	Al, Zr	...	...
Raises solvus temperature of γ'	...	...	Co
Lowers solvus temperature of γ'	...	...	Cr
Hardening precipitates and/or intermetallics	Al, Ti, Nb	Al, Mo, Ti(b), W, Ta	Al, Ti, Nb
Oxidation resistance	Cr	Al, Cr, Ta	Al, Cr, Ta
Improves hot corrosion resistance	La, Y	La, Y, Th	La, Th
Sulfidation resistance	Cr	Cr	Cr
Enhances creep-rupture properties by grain-boundary morphology changes	B	B, Zr	B(c), Zr
Improves intermediate-temperature ductility	...	...	Hf
Causes grain-boundary segregation	...	...	B, C, Zr
Facilitates working	...	Ni_3Ti	...

(a) Not all of these effects necessarily occur in a given alloy. (b) Hardening by precipitation of Ni_3Ti also occurs if sufficient Ni is present. (c) If present in large amounts, borides are formed.

Reduction in chromium causes a decrease in hot corrosion resistance so that superalloys of this class containing about 8 to 12% Cr may require the use of coatings, such as diffused aluminum, to compensate for the loss in hot corrosion resistance. Addition of small amounts of hafnium (1 to 1.5%) causes a marked improvement in the intermediate-temperature ductility of these nickel superalloy castings. Because of this high-temperature strength advantage, many aircraft gas turbine engines use nickel γ'-strengthened superalloy castings for high-stress, high-temperature turbine blade applications. The fine-grain structure of forgings favors low-cycle fatigue strength at low-to-intermediate temperatures, hence the use of forgings in disk applications.

It is obvious that superalloys are complex compositional combinations and that various elements have significant effects on properties. The alloying elements in superalloys can be classified as: (*a*) solid-solution strengtheners that affect only the matrix; (*b*) elements that form carbides or intermetallic compounds in grain boundaries or within

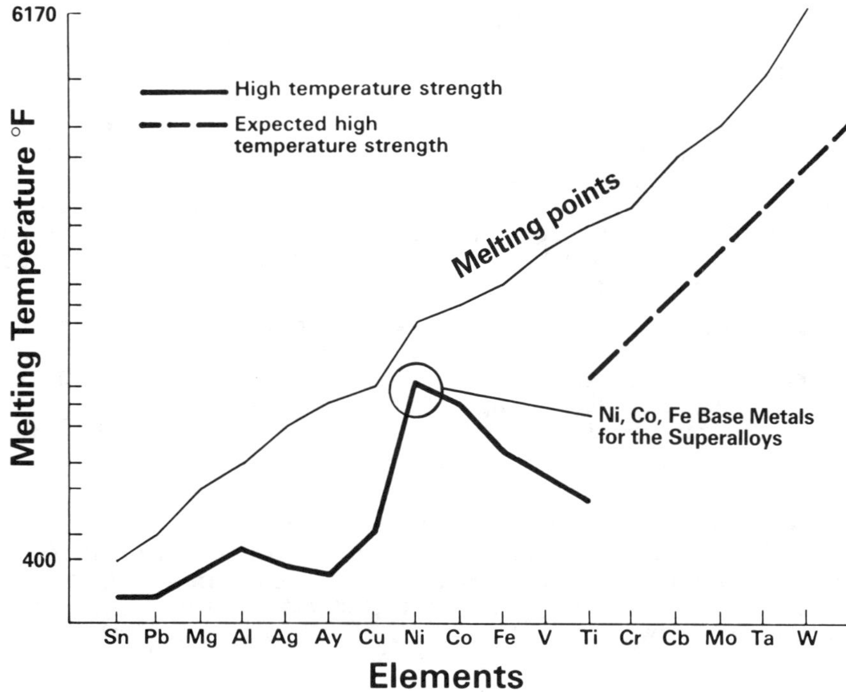

Fig. 2.1 Melting characteristics of the elements vs high-temperature strength

grains; (*c*) elements that form a coherent γ' phase in nickel-base alloys; and (*d*) elements that have miscellaneous effects, such as improving oxidation resistance, raising or lowering the γ' solid solubility temperature, or improving intermediate-temperature ductility. Table 2.3 briefly reviews the alloying effects of various elements on superalloys.

PRINCIPLES OF HIGH-TEMPERATURE BEHAVIOR

The high-temperature strength behavior of the metallic elements approximately follows their melting characteristics, i.e., the higher the melting point, the greater the high-temperature strength of the metal. Close examination of this relationship, however, indicates that some elements fall off the curve, notably titanium. Nonetheless, the melting point of the elements is an important property and explains the role of iron, nickel, and cobalt as base metals for superalloys (Fig. 2.1).

Creep/Stress-Rupture

In service at elevated temperatures, the life of a metal component subjected to either static or dynamic loading is predictably limited. In contrast, at lower temperatures, and in the absence of a corrosive environment, the life of a part is unlimited under static conditions, provided the operational loads do not exceed the yield strength of the metal. Stress imposed at elevated temperatures, however, produces a continuous strain in the component and results in creep. By definition, creep is time-dependent strain or deformation occurring under stress at elevated temperature. After a period of time, creep terminates in fracture called "stress-rupture." The conditions of temperature, stress, and time under which creep and stress-rupture failures occur depend on the metal and the service environment. Consequently, elevated temperature failures may occur over a wide range of temperatures.

Creep generally occurs at a temperature slightly above the recrystallization temperature of the metal involved, when the atoms become sufficiently mobile to allow time-dependent rearrangement of structure. The elevated temperature behavior of a part - the mechanical strength of the metal becoming limited by creep rather than by yield strength - must be determined for each material on the basis of individual characteristics.

Parts can fail at elevated temperatures for many reasons other than creep and stress-rupture, such as low-cycle and high-cycle fatigue, thermal fatigue, tension overload, and combinations of these conditions. However, it is the creep/stress-rupture behavior that differentiates between low-temperature and high-temperature conditions.

The primary metallurgical factor in stress-rupture behavior is the transition from transgranular to intergranular fracture, another important behavior difference between low and high temperatures. At low temperatures, the grain-boundary regions are stronger than grains, and thus, deformation and fracture are transgranular. At high temperatures, grain boundaries are weaker than grains, and deformation and fracture are largely intergranular. The temperature at which this transition occurs is called the "equicohesive" temperature (Fig. 2.2). This temperature varies with exposure time and stress. For each combination of stress and rupture life, there is a temperature above which all stress-rupture fractures will be intergranular. For shorter rupture lives and lower stresses, this transgranular-to-intergranular transition will occur at higher temperatures. For longer rupture lives and higher stresses, this transition will occur at lower temperatures. Under certain conditions, features of both transgranular and intergranular fracture will be formed in stress-rupture failures. Thus, the equicohesive transition temperature is variable, depending on time and stress conditions.

Metallurgical Instabilities

Another characteristic of high-temperature service involves metallurgical instability. Stress, time, temperature, and environment may act to change the metallurgical structure during operation and, thereby, contribute to failure by reducing strength and/or ductility. It should be noted that in some few cases strength may be enhanced. These structural changes or metallurgical instabilities are best described in terms of their influence on stress-rupture properties. A sharp change downward in the slope of the stress-rupture curve indicates that failures will occur in shorter times and at lower stresses than

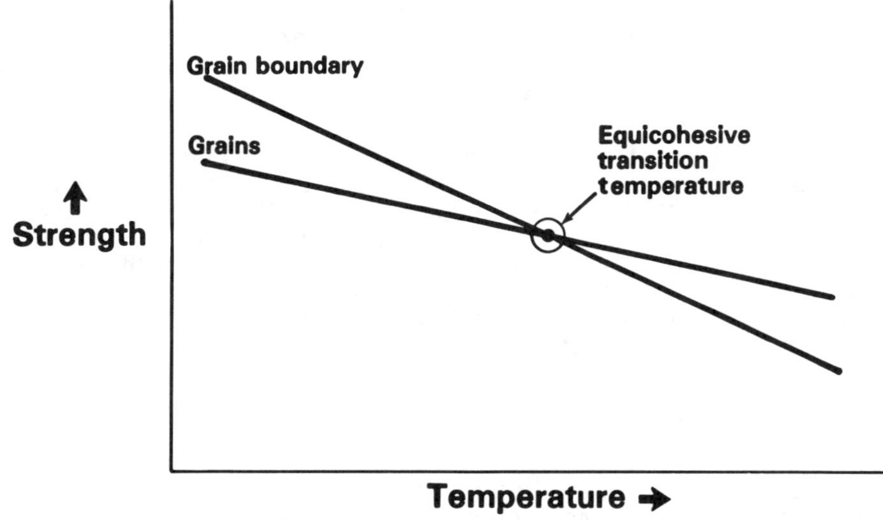

Fig. 2.2 Determination of equicohesive transition temperature

originally predicted. Instabilities usually are associated with aging (phase precipitation), overaging (phase coalescence and coarsening), phase decomposition (generally involving carbides, borides, and nitrides), intermetallic phase precipitation, order-disorder transition, internal oxidation, and stress-corrosion. Typical instability problems with the γ' nickel-base superalloys involve intermetallic phase precipitation. Close-packed sigma (σ), mu (μ), and Laves phases form at elevated temperatures with generally deteriorating effects on stress-rupture properties.

Another instability characteristic involves carbide reactions. A variety of carbide types are found in superalloys. Although temperature and stress affect both carbides within grains and carbides at grain boundaries, the effects on grain-boundary carbides usually are a much more significant factor in altering creep and rupture behavior. Grain-boundary morphology is indeed important relative to high-temperature properties. The presence of carbides at grain boundaries as strengtheners is necessary for optimum creep and rupture life, but alteration in shape or breakdown to other carbide forms may cause property degradation. The best type of carbide formation at grain boundaries for optimum strength is discrete, blocky particles. Acicular carbides at grain boundaries do not act as brittle notch formers to adversely affect rupture life, but they do tend to reduce impact strength. Continuous carbide films at the boundaries, on the other hand, substantially reduce stress-rupture life.

The carbide forms in superalloys are classified as MC, $M_{23}C_6$, M_6C, and Cr_7C_3. The "M" constituent in these carbides usually is titanium, but other elements such as molybdenum, niobium, vanadium, zirconium, and tantalum also may be involved. The Cr, of course, in Cr_7C_3 is chromium. It is necessary through proper processing and heat treatment to maintain these various carbides at the grain boundaries as discontinuous, blocky particles to develop optimum stress-rupture properties. Continuous-film and fine, cellular carbides are to be avoided for optimum ductility and rupture life. Small additions of boron and zirconium are essential in nickel superalloys to enhance creep-rupture properties by strengthening grain boundaries.

An understanding of the presence of tramp elements in nickel-base superalloys is necessary to completely grasp the effect of composition on superalloy properties. Although not actually an instability, the presence of some elements can cause significant reduction in properties. Minute quantities of lead, antimony, bismuth, selenium, arsenic, or silver, measured in parts-per-million quantities, can cause significant loss in stress-rupture life and ductility, with devastating effects on the high-temperature service life of parts.

STRENGTHENING MECHANISMS

Superalloys are complex alloy systems that utilize numerous strengthening mechanisms, including the following:

* Solid-solution strengthened γ
* Increased volume percentage of γ' or γ''
* Solid-solution hardened γ'
* Minimal formation of σ, μ, and Laves phases
* Control of carbides to prevent denuded zones, $M_{23}C_6$ grain-boundary films, and Widmanstätten M_6C for tensile strength
* Control of carbides and grain-boundary γ' to enhance rupture strength
* Control of component thickness to grain size ratio

Solid-solution strengthening is common to all superalloys and occurs, as previously defined, when the matrix or base background metal dissolves an alloying addition in the solid state with a stiffening effect.

The most dramatic and effective of the strengthening mechanisms is the precipitation of an intermetallic phase from solid solution. Relative to superalloys, formation of the γ' phase $Ni_3(Al,Ti)$ in nickel-base alloys is by far the most important. To develop such phases, a supersaturated solid solution (a solution holding more dissolved material than normally possible) is formed first, which later precipitates out the excess material. A supersaturated solid solution at lower temperature is created by rapidly cooling from the higher temperature solid solubility area.

Under these conditions, the lower temperature solid solution will contain more of the second material - in the nickel superalloys, the $Ni_3(Al,Ti)$ intermetallic - in solution than normally possible, thus forming the supersaturated condition and thus providing the mechanism for precipitation of the strengthening phase. The ternary alloys Ni-Cr-Al and Ni-Cr-Ti precipitate the intermetallics Ni_3Al and Ni_3Ti when the solubility limits are exceeded (Fig. 2.3 and 2.4). Figure 2.5 illustrates the supersaturated solid solution and precipitation concepts in the binary aluminum-copper system.

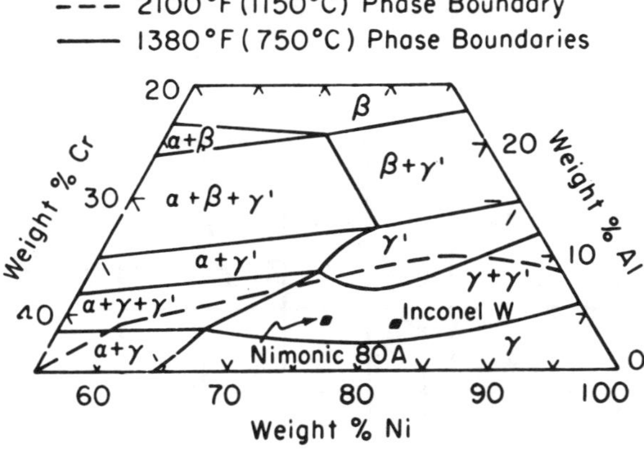

Fig. 2.3 Isothermal section of the nickel-rich portion of the nickel-chromium-aluminum system

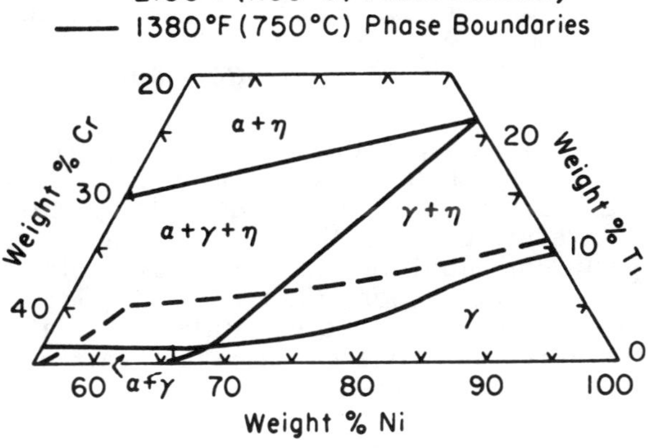

Fig. 2.4 Isothermal section of the nickel-rich portion of the nickel-chromium-titanium system

As stated above, the γ′ precipitate $Ni_3(Al,Ti)$ is the most important of the superalloy strengthening mechanisms. As such, increasing the volume percentage of γ′ improves high-temperature strength. It is not surprising, therefore, that increasing the aluminum and titanium contents in nickel-base superalloys increases high-temperature strength. Chromium, which is so important for corrosion and oxidation resistance, has a negative effect on high-temperature strength, i.e., an increase in chromium causes a decrease in strength. This is due to the fact that as chromium is increased the γ′ solid solubility temperature decreases, thereby decreasing second-phase strengthening (Fig. 2.6).

Another important strengthening mechanism in nickel-base superalloys is the formation of the Ni_3Nb intermetallic. This phase, known as gamma double prime (γ″), exhibits a body-centered tetragonal (bct) structure rather than the fcc structure of the $Ni_3(Al,Ti)$ γ′ phase or

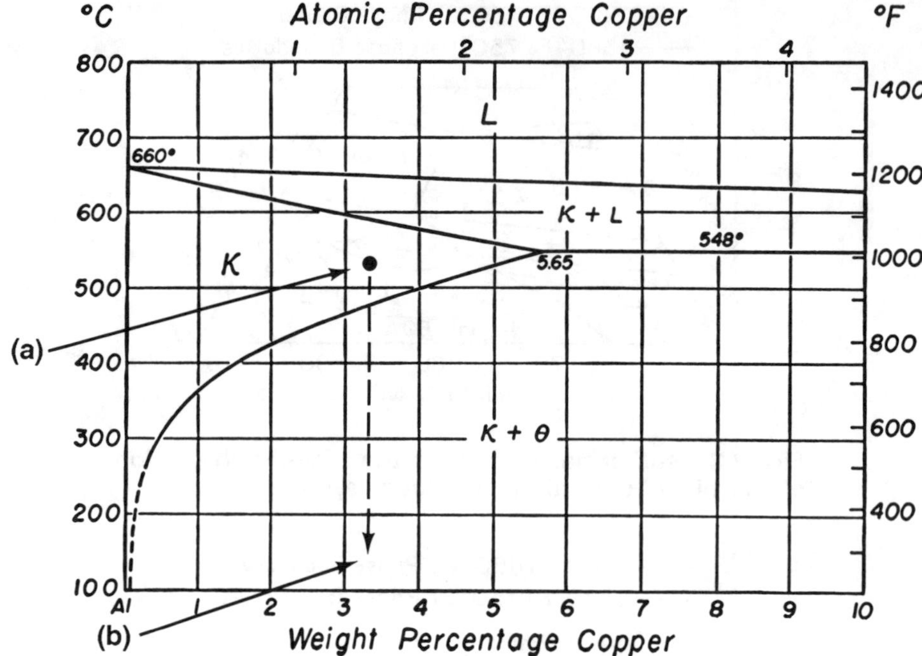

(a) At this temperature, the solid solution of aluminum naturally contains 3.5% Cu. (b) After rapid cooling to room temperature, supersaturated solution is created with aluminum holding 3.5% Cu in solution, which is much more than normally possible. This excess will precipitate out as a strengthening phase during intermediate-temperature aging.

Fig. 2.5 Supersaturated solid solution and precipitation concepts, as illustrated by the aluminum-copper binary system

the γ austenitic matrix phase. There are four important nickel superalloys containing iron (INCO 718, INCO 706, Rene′ 62, and Udimet 630) that depend primarily on γ'' precipitation strengthening instead of, or in conjunction with, γ' strengthening. These alloys contain 2 to 6% Nb and considerable amounts of iron. Of the four alloys, INCO 718 is the most widely used – indeed a workhorse alloy.

Dispersion strengthening is another mechanism of particular importance for very high temperature operation. This mechanism utilizes the presence of a dispersed phase in the structure in place of, or in addition to, precipitated phases. These constituents, such as yttrium oxide or thorium dioxide, differ from the carbide, γ', or γ'' phases in that they persist in the structure throughout the solid state up to the melting point. In the case of the dispersed phase, there is no solid solubility temperature that distinguishes it from the other phases. Obviously, the presence of a strengthening phase at these high temperatures improves strength. A number of alloys, identified by

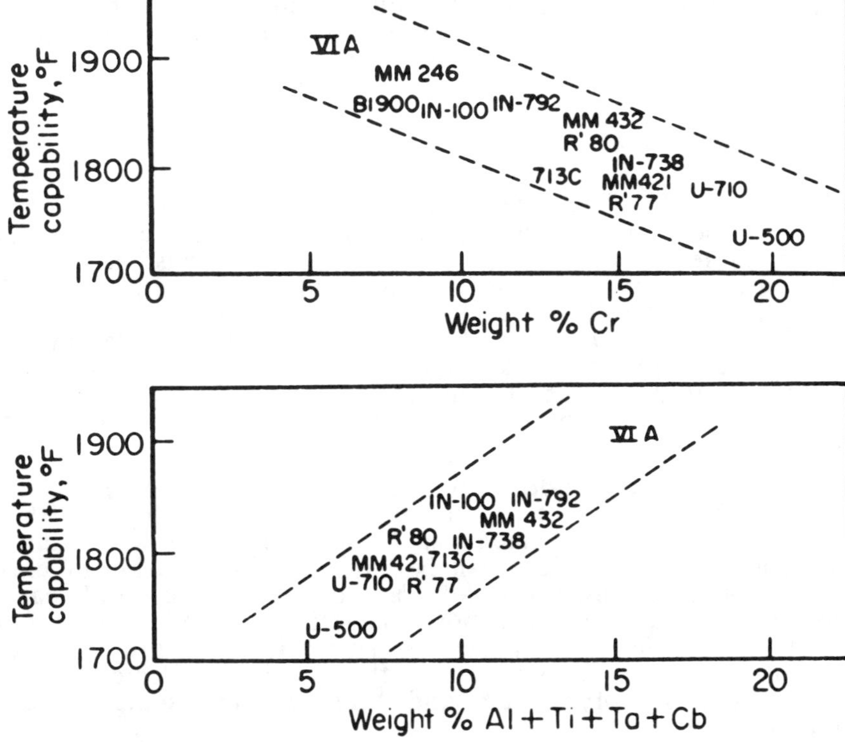

Fig. 2.6 Effect of Cr+Al+Ti+Nb in nickel-base superalloys

the designation "MA" (mechanical alloying), have been developed by INCO Limited and utilize P/M processing and dispersion strengthening techniques.

SUPERALLOY CLASSIFICATIONS

As previously stated, relative to high-temperature strength, the nickel-base precipitation-strengthened superalloys are outstanding. At very high temperatures, it is true, the oxide-dispersion-strengthened alloys and even some cobalt alloys exhibit superior strength to the precipitation-strengthened nickel superalloys. (See Fig. 1.2 in Chapter 1.) However, both of these alloy types are much weaker than the nickel superalloys at somewhat lower temperatures. The iron-base superalloys and the solid-solution-strengthened alloys are the weakest of the superalloys at elevated temperatures. Because of these strength characteristics, the iron-base superalloys are being used in fewer applications as service temperature trends increase upward. This is not true for the nickel-base, iron-containing INCO 718 alloy, which continues to be used widely. The main advantage of iron-base superalloys is lost. Solid-solution alloys, of course, continue to be used when strength requirements are not important, but oxidation and hot corrosion resistance as well as fabricability are required. Use of cobalt-base superalloys has decreased, largely because of post-cartel action with the metal. Cobalt alloys such as MAR-M 509 are used widely for aircraft engine turbine vane castings, and Haynes 188 cobalt alloy is a popular combustion material.

Iron-Base Superalloys

Iron-base superalloys are defined as those alloys that have iron as the major constituent, with significant amounts of chromium and nickel and possibly lesser amounts of molybdenum or tungsten. They also are strengthened by a carbide or intermetallic precipitate and/or solid solution. The intermetallic precipitate is usually of the $Ni_3(Al,Ti)$ γ' type. They differ from stainless steels in their chromium-nickel ratios and strengthening mechanisms. The stainless steels, of course, contain 12 to 25% Cr and 0 to 20% Ni. The iron-base superalloys contain more than 20% Ni (usually from 25 to 35%).

Various elements are added to perform one or more desirable functions. The most potent strengthening for alloys with fcc matrixes is provided by elements such as nickel, aluminum, titanium, and niobium. The fcc alloys can be hardened by addition of carbon in relatively large amounts (~0.5%) to form a general carbide precipitate; nitrogen and phosphorus sometimes are added to enhance this effect. Carbon also promotes formation of grain-boundary carbide phases to provide strength in these regions. Elements in solid solution, such as molybdenum and tungsten, add some degree of strengthening.

Oxidation resistance principally is provided by the use of chromium. Nickel and manganese also improve oxidation resistance. Small additions of boron greatly improve elevated temperature properties. Iron-base alloys of most importance for application temperatures above 540 °C (1000 °F) are those with an fcc matrix, because a close-packed lattice is more resistant to time-dependent deformation processes.

Iron-base superalloys strengthened by intermetallic compound precipitation have found primary usage in gas turbine engines as blades, disks, castings, and fasteners. For example, some gas turbine engines use A-286 for turbine disks and hubs. A-286 also has been used for turbine cases. Alloys relying solely on precipitation of carbides have not been used widely in the United States.

There are many superalloys containing significant amounts of iron that do not truly fall into the iron-base category, because their compositions are complex combinations of iron, nickel, chromium, and possibly cobalt, with lesser amounts of molybdenum, tungsten, or niobium on occasion. These alloys also are strengthened by solid solution and/or precipitation of intermetallics. In some instances, the iron-nickel ratios are such that the alloys should be classified as nickel base. Examples of this are the solid-solution-strengthened Hastelloy X containing 16% Fe and 49% Ni and the γ''-strengthened INCO 718 alloy containing 18.5% Fe and 52.5% Ni. In these two examples, the alloys are properly classified as nickel-base superalloys containing iron as an alloy addition.

Other alloys of this type are not so clearly defined. For example, the γ'-strengthened INCO 901 superalloy contains 42.5% Ni and 36% Fe.

This alloy could be considered nickel base, or it could fall into the complex iron-nickel-chromium category. An example of a complex composition solid-solution-strengthened alloy is Multimet (or N-155), which contains 21% Cr, 20% Ni, 20% Co, 32.5% Fe, 3% Mo, 2.5% W, and 1% Nb.

Another complex alloy of this class is Incoloy 903, a precipitation-strengthened iron-nickel-cobalt alloy containing 41% Fe, 38% Ni, 15% Co, 1% Al, 2% Ti, and 3% Nb. This alloy is used for its constant low coefficient of thermal expansion in addition to its high strength. This characteristic is useful in applications requiring close operating clearance over a range of temperatures.

Cobalt-Base Superalloys

Cobalt-base superalloys are defined as those alloys that have cobalt as the major constituent, with significant amounts of nickel, chromium, and tungsten and lesser amounts of molybdenum, niobium, tantalum, titanium, lanthanum, and iron on occasion. They are strengthened by solid-solution and carbide phases. Those alloys that depend primarily on carbide-phase strengthening contain 0.4 to 0.85% carbon.

Cobalt solid-solution alloys can be subdivided into three groups on the basis of use: (*a*) alloys for use primarily at temperatures from 650 to 1150 °C (1200 to 2100 °F), including Haynes 25, Haynes 188, UMCo-50, and S-816; (*b*) fastener alloys MP-35N and MP-159, for use to about 650 °C (1200 °F); and (*c*) wear-resistant Stellite 6B.

All alloys in the heat treated and softened condition have fcc crystal structures; however, alloys MP-35N and MP-159 develop controlled amounts of close-packed hexagonal (cph) structure during the thermomechanical processing that is recommended before service applications. Stellite 6B heat treated between 650 and 1050 °C (1200 and 1900 °F) and Haynes 25 exposed for 1000 h or more at temperatures near 650 °C (1200 °F) may partly transform to a cph structure.

None of the cobalt-base superalloys is a complete solid-solution alloy, because all contain secondary carbide phases, or intermetallic compounds. Aging causes additional secondary phase precipitation, which generally results in loss of some room-temperature ductility.

Haynes 25 is a well-known wrought cobalt-base alloy that has been widely used for hot sections of gas turbines, for nuclear reactor components, for surgical implants, and in the cold worked condition, for fasteners and wear pads.

Haynes 188 is specially designed for sheet metal components, such as combustors and transition ducts, in gas turbines. The basic composition, provided that lanthanum, silicon, aluminum, and manganese contents are judiciously controlled, provides excellent qualities such as oxidation resistance at temperatures up to 1100 °C (2000 °F), hot corrosion resistance, creep resistance, room-temperature formability, and ductility after long-term aging at service temperatures.

UMCo-50, which contains about 21% Fe, is not as strong as Haynes 25 or Haynes 188. It is not used extensively in the United States, particularly not in gas turbine applications. In Europe, on the other hand, it is used extensively for furnace parts and fixtures.

MP-35N and MP-159 were designed specifically to be work hardened, and both alloys have high strength and relatively high ductility in the work-hardened condition. The combination of high strength and high ductility in these alloys is attributed to the formation of small platelets of cph structure in the work-hardened fcc matrix. MP-35N and MP-159 are used predominantly for fasteners; MP-159 has an approximate upper limit on service temperature of 650 °C (1200 °F).

The last group of high-temperature, solid-solution-strengthened cobalt alloys consists of a single alloy, Stellite 6B. This alloy is characterized by high hot hardness and relatively good resistance to oxidation. The latter property is derived chiefly from the high chromium content (about 30%), whereas the hot hardness is obtained through formation of complex carbides. Stellite 6B is used widely for erosion shields in steam turbines, for wear pads in gas turbines, and for bends in tube systems carrying particulate matter at high temperatures and high velocities.

Carbide-phase-strengthened cobalt superalloys such as X-40, WI-52, MAR-M 302, and MAR-M 509 (all containing high carbon contents) are used predominantly for turbine engine airfoils - primarily for static vane applications. These alloys have sufficiently good high-temperature strength, oxidation resistance, and good repairability by welding for this application.

Nickel-Base Superalloys

Nickel-base superalloys are defined as those alloys that have nickel as the major constituent, with significant amounts of chromium. They may contain cobalt, iron, molybdenum, tungsten, and tantalum as major alloy additions. They are strengthened by solid-solution and second-phase intermetallic precipitation. The intermetallic forming elements are aluminum, titanium, and niobium.

Nickel-base superalloys contain 30 to 75% Ni and up to 30% Cr. Iron contents range from relatively small amounts in most Inconels, Nimonics, and Hastelloys to about 35% in alloys such as Incoloy 901 and Inconel 706. Many nickel-base alloys contain small amounts of aluminum, titanium, niobium, molybdenum, and tungsten to enhance either strength or corrosion resistance.

The combination of nickel and chromium gives these alloys outstanding oxidation resistance. As a class, nickel-base superalloys exceed stainless steels in mechanical strength, especially at temperatures above 650 °C (1200 °F).

Nickel-base superalloys are used in a wide range of applications that require high strength or resistance to oxidation and corrosion. Solid-solution alloys such as Inconel 600, Inconel 601, and RA 333 are used frequently for furnace parts and other heat treating equipment. These alloys also are used in high-temperature chemical processing equipment such as hydrocarbon reformers and crackers.

Power generation is another field in which nickel-base superalloys are used extensively. In nuclear power plants, applications include steam generator tubing and structural components of reactor cores.

In fossil-fueled plants, these alloys are used for superheater tubing, ash-handling systems, stack-gas scrubbers, and other parts that must resist heat or corrosion, or both.

Solid-solution nickel alloys normally are used in the annealed temper, the annealing treatment depending on the properties required by the application. In general, a relatively low annealing temperature of 870 to 980 °C (1600 to 1800 °F) is used to produce the highest tensile and fatigue strengths. A high-temperature anneal at about 1120 to 1200 °C (2050 to 2200 °F) usually produces optimum fatigue resistance and creep-rupture properties at service temperatures greater than 600 °C (1100 °F).

Some solid-solution nickel alloys, notably Hastelloy X, Inconel 601, Inconel 617, and Inconel 625, are used in aerospace applications. Hastelloy X is used widely for turbine engine combustors, whereas Inconel 601 is used for jet engine igniters and for gas turbine components, such as combustion can liners and diffuser assemblies. The high stress-rupture strength of Inconel 617 at temperatures higher than 980 °C (1800 °F) makes it a useful material for various components of aircraft gas turbines. Inconel 625 is used for heat shields, aircraft ducting systems, exhaust systems, thrust reversers, fuel and hydraulic lines, spray bars, and turbine shroud rings.

Precipitation-strengthening nickel alloys contain aluminum, titanium, or niobium to cause precipitation of a second phase during appropriate heat treatment. The precipitated phase, usually γ' or γ'', substantially increases the strength and hardness of the alloy. For example, Inconel X-750, a precipitation-strengthening version of Inconel 600, has a yield strength about three times that of Inconel 600 at 540 °C (1000 °F). The precipitation-strengthening alloy Nimonic 80A shows a similar increase over its solid-solution counterpart, Nimonic 75.

Heat treatments for the precipitation-strengthening alloys generally consist of a solution treatment at 970 to 1175 °C (1700 to 2150 °F) followed by one or more precipitation treatments at 600 to 815 °C (1100 to 1500 °F).

Most of these alloys utilize aluminum and titanium to promote precipitation strengthening by forming the γ' intermetallic $Ni_3(Al,Ti)$. In some alloys, niobium is used along with lesser amounts of aluminum and titanium, thus forming the γ'' intermetallic Ni_3Nb. The niobium-strengthened alloys, Inconel 718 for example, exhibit delayed responses to precipitation-hardening temperatures. Such alloys have significantly better weldability, because the heat of welding does not induce hardening and consequent postweld cracking.

Aerospace applications constitute the greatest field of use for precipitation-strengthening superalloys. They are used in rocket engines and for such aircraft gas turbine parts as blades, disks, rings, shafts, and various compressor and diffuser components. Other applications include bolts and springs in nuclear reactors.

Applications for precipitation-strengthening alloys frequently involve forged components. Consequently, mechanical properties in the forged-plus-heat treated condition are often important. Also, because statistical methods frequently must be used to determine the reliability of a given part, statistical distributions of properties are of considerable help in designing parts, particularly critical aircraft parts.

Because castings are intrinsically stronger than forgings at elevated temperatures, investment cast nickel superalloys are used for applications involving the highest temperature service conditions. Thus, cast nickel superalloys such as INCO 713, INCO 100, B-1900, MAR-M 247, and MAR-M 200 are used for turbine blades. In the most severe applications for these critical rotating parts, directionally solidified structures or single-crystal components are selected. In high-strength cast alloys, such as B-1900, MAR-M 247, and MAR-M 200, small amounts of hafnium are added to improve intermediate-temperature ductility.

CONCLUSION

Use of superalloys has expanded beyond aircraft gas turbine engines into many other industries, but aircraft engines will continue to be the most important market as superalloys remain the cornerstone high-temperature material class for this application. Because of higher engine operating temperatures, the amount of superalloys used expressed as a percentage of total engine weight has increased from about 20% to approximately 50%. This trend will continue to reach about 60% in the 1990's.

Nickel-base γ'-strengthened superalloys are predominant, particularly for aircraft engine use. However, INCO 718 alloy, strengthened by the γ'' intermetallic precipitate (Ni_3Nb), has become and will continue to be the workhorse alloy for aircraft engines. Nickel-base precipitation-strengthened superalloys containing iron, with the exception of INCO 718, will not be used as much as the iron-free alloys. Also, use of the precipitation-strengthened iron-base and complex iron-nickel-chromium alloys is declining. Solid-solution-strengthened nickel superalloys like Hastelloy X containing chromium, iron, and molybdenum are used widely.

Use of cobalt-base superalloys is minimal. Cobalt alloys such as MAR-M 509 are used for aircraft turbine engine airfoils, particularly the static vanes, and Haynes 188 cobalt-base alloy remains a popular combustor alloy. Cobalt alloys also are used for medical implant applications.

New superalloy compositions have not been plentiful, but a great number of varying compositions are already available. INCO's mechanical alloyed products, involving oxide dispersion strengthening, have gained some small acceptance with General Electric's production use of MA 754 alloy for military engine turbine vanes. There has been some activity in new alloy compositions for P/M applications, such as Pratt & Whitney's MERL 76 alloy, which is very similar to IN-100 alloy. Pratt & Whitney also has developed a new single-crystal alloy characterized by the absence of carbon, boron, zirconium, and hafnium (all grain-boundary elements) and with 12% tantalum.

REFERENCES

1. *The Nimonic Alloys, And Other Nickel-Base High-Temperature Alloys,* edited by W. Betteridge and J. Heslop, Crane, Russak, New York, 1974.

2. *The Superalloys*, edited by C.T. Sims and W.C. Hagel, John Wiley & Sons, New York, 1972.

3. *Source Book on Materials for Elevated-Temperature Applications,* edited by E.F. Bradley, American Society for Metals, Metals Park, OH, 1979.

Chapter 3

Microstructure

PHASES AND STRUCTURES OF SUPERALLOYS

Superalloys consist of the austenitic face-centered cubic (fcc) matrix phase gamma (γ) plus a variety of secondary phases. The principal secondary phases are the carbides MC, $M_{23}C_6$, M_6C, and M_7C_3 (rare) in all superalloy types and gamma prime (γ') fcc ordered $Ni_3(Al,Ti)$ intermetallic compound in nickel and iron-nickel superalloys. In some alloys, the gamma double prime (γ'') Ni_3Nb compound is present. Superalloys derive their strength from solid-solution hardeners and precipitating phases. Carbides may provide limited strengthening directly (e.g., through second phase strengthening) or, more commonly, indirectly (e.g., by stabilizing grain boundaries against excessive shear). In addition to those elements that produce solid-solution strengthening and promote carbide and γ' formation, other elements (e.g., boron, zirconium, hafnium, and cerium) are added to enhance mechanical or chemical properties. Some carbide- and γ'-forming elements may contribute significantly to chemical properties as well. Table 3.1 lists the secondary phases that may be found in superalloy matrix structures. Tables 3.2, 3.3, and 3.4 give typical composition ranges for select classes of heat-resistant superalloys.

The principal microstructural variables of superalloys are: (*a*) the precipitate amount and its morphology; (*b*) grain size and shape; and (*c*) carbide distribution. Nickel and iron-nickel superalloy properties are controlled by all three variables; the first variable is essentially absent in cobalt-base superalloys. Structure control is achieved through composition selection/modification and by processing. For a given nominal composition, there are property advantages and disadvantages of the structures produced by deformation processing and by casting. Cast superalloys generally have coarser grain sizes, more alloy segregation, and improved creep and rupture characteristics. Wrought superalloys generally have more uniform, and usually finer, grain sizes and improved tensile and fatigue properties.

Nickel and iron-nickel superalloys typically consist of γ' dispersed in a γ matrix, and the strength increases with increasing the volume fraction (V_f) of the γ'. The lowest volume fraction amounts of γ' are found in iron-nickel superalloys and first-generation nickel-base superalloys, where volume fraction amounts of γ' are generally less than about 0.25 (25 vol%). The γ' is commonly spheroidal in lower volume fraction γ' alloys, but often cuboidal in higher volume

Table 3.1 Constituents commonly observed in superalloys

Phase	Crystal structure	Formula	Comments
γ'	fcc	Ni_3Al $Ni_3(Al,Ti)$	Principal strengthening phase in many nickel- and nickel-iron-base superalloys; crystal lattice varies slightly in size (0 to 0.5%) from that of austenite matrix; shape varies from spherical to cubic; size varies with exposure time and temperature
η	hcp	Ni_3Ti (no solubility for other elements)	Found in iron-, cobalt-, and nickel-base superalloys with high titanium/aluminum ratios after extended exposure; may form intergranularly in a cellular form or intragranularly as acicular platelets in a Widmanstätten pattern
γ''	bct	Ni_3Nb	Principal strengthening phase in Inconel 718; γ'' precipitates are coherent disk-shaped particles that form on the {100} planes (avg diam approximately 600 Å, thickness approximately 50 to 90 Å); metastable phase
Ni_3Nb (δ)	orthorhombic	Ni_3Nb	Observed in overaged Inconel 718; has an acicular shape when formed between 815 and 980 °C (1500 and 1800 °F); forms by cellular reaction at low aging temperatures and by intragranular precipitation at high aging temperatures
MC	cubic	TiC NbC HfC	Titanium carbide has some solubility for nitrogen, zirconium, and molybdenum; composition is variable; appears as globular, irregularly shaped particles that are gray to lavender; "M" elements can be titanium, tantalum, niobium, hafnium, thorium, or zirconium
$M_{23}C_6$	fcc	$Cr_{23}C_6$ $(Cr,Fe,W,Mo)_{23}C_6$	Form of precipitation is important; it can precipitate as films, globules, platelets, lamellae, and cells; usually forms at grain boundaries; "M" element is usually chromium, but nickel-cobalt, iron, molybdenum, and tungsten can substitute
M_6C	fcc	Fe_3Mo_3C Fe_3W_3C–Fe_4W_2C Fe_3NB_3C Nb_3Co_3C Ta_3Co_3C	Randomly distributed carbide; may appear pinkish; "M" elements are generally molybdenum or tungsten; there is some solubility for chromium, nickel-niobium, tantalum, and cobalt
M_7C_3	hexagonal	Cr_7C_3	Generally observed as a blocky intergranular shape; observed only in alloys such as Nimonic 80A after exposure above 1000 °C (1830 °F), and in some cobalt-base alloys
M_3B_2	tetragonal	Ta_3B_2 V_3B_2 Nb_3B_2 $(Mo,Ti,Cr,Ni,Fe)_3B_2$ Mo_2FeB_2	Observed in iron-nickel- and nickel-base alloys with about 0.03% B or greater; borides appear similar to carbides, but are not attacked by preferential carbide etchants; "M" elements can be molybdenum, tantalum, niobium, nickel, iron, or vanadium
MN	cubic	TiN (Ti,Nb,Zr)N (Ti,Nb,Zr) (C,N) ZrN NbN	Nitrides are observed in alloys containing titanium, niobium, or zirconium; they are insoluble at temperatures below the melting point; easily recognized as-polished, having square to rectangular shapes and ranging from yellow to orange
μ	rhombohedral	Co_7W_6 $(Fe,Co)_7(Mo,W)_6$	Generally observed in alloys with high levels of molybdenum or tungsten; appears as coarse, irregular Widmanstätten platelets; forms at high temperatures
Laves	hexagonal	Fe_2Nb Fe_2Ti Fe_2Mo Co_2Ta Co_2Ti	Most common in iron-base and cobalt-base superalloys; usually appears as irregularly shaped globules, often elongated, or as platelets after extended high-temperature exposure
σ	tetragonal	FeCr FeCrMo CrFeMoNi CrCo CrNiMo	Most often observed in iron- and cobalt-base superalloys, less commonly in nickel-base alloys; appears as irregularly shaped globules, often elongated; forms after extended exposure between 540 and 980 °C (1005 to 1795 °F)

Table 3.2 Nominal compositions of wrought heat-resistant alloys

Alloy	UNS number	Cr	Ni	Co	Mo	W	Nb	Ti	Al	Fe	C	Other
		Composition, %										
Iron-base solid-solution alloys												
17-14CuMo	...	16.0	14.0	...	2.50	...	0.4	0.3	...	62.4	0.12	0.75 Mn; 0.50 Si; 3.0 Cu
19-9DL	K63198	19.0	9.0	...	1.25	1.25	0.4	0.3	...	66.8	0.30	1.10 Mn; 0.60 Si
Carpenter 20Cb-3	N08020	20.0	34.0	...	2.50	...	1.0 max	...	...	42.4	0.07 max	3.5 Cu
Incoloy 800	N08800	21.0	32.5	...	...	...	...	0.38	0.38	45.7	0.05	...
Incoloy 801	N08801	20.5	32.0	...	...	...	...	1.13	...	46.3	0.05	...
Incoloy 802	...	21.0	32.5	...	...	...	...	0.75	0.58	44.8	0.35	...
N-155	R30155	21.0	20.0	20.0	3.00	2.5	1.0	...	...	32.2	0.15	0.15 N; 0.02 La; 0.02 Zr
RA330	N08330	19.0	36.0	...	...	...	...	...	...	45.1	0.05	...
Cobalt-base solid-solution alloys												
Haynes 25 (L-605)	R30605	20.0	10.0	50.0	...	15.0	...	...	...	3.0	0.10	1.5 Mn
Haynes 188	R30188	22.0	22.0	37.0	...	14.5	...	...	...	3.0 max	0.10	0.90 La
S-816	R30816	20.0	20.0	42.0	4.0	4.0	4.0	...	...	4.0	0.38	...
Stellite 6B	...	30.0	1.0	61.5	...	4.5	...	...	...	1.0	1.0	...
UMCo-50	...	28.0	...	49.0	...	...	...	...	...	21.0	0.12 max	...
Nickel-base solid-solution alloys												
Hastelloy B	N10001	1.0 max	63.0	2.5 max	28.0	...	...	...	...	5.0	0.05 max	0.03 V
Hastelloy B-2	N10665	1.0 max	69.0	1.0 max	28.0	...	...	...	...	2.0 max	0.02 max	...
Hastelloy C	N10002	16.5	56.0	...	17.0	4.5	...	...	...	6.0	0.15 max	...
Hastelloy C-4	N06455	16.0	63.0	2.0 max	15.5	...	...	0.7 max	...	3.0 max	0.015 max	...
Hastelloy C-276	N10276	15.5	59.0	...	16.0	3.7	...	...	...	5.0	0.02 max	...
Hastelloy N	N10003	7.0	72.0	...	16.0	...	...	0.5 max	...	5.0 max	0.06	...
Hastelloy S	...	15.5	67.0	...	15.5	...	...	...	0.2	1.0	0.02 max	0.02 La
Hastelloy W	N10004	5.0	61.0	2.5 max	24.5	...	...	...	...	5.5	0.12 max	0.6 V
Hastelloy X	N06002	22.0	49.0	1.5 max	9.0	0.6	...	...	2.0	15.8	0.15	...
Inconel 600	N06600	15.5	76.0	...	...	...	...	...	...	8.0	0.08	0.25 max Cu
Inconel 601	N06601	23.0	60.5	...	...	...	...	...	1.35	14.1	0.05	0.5 max Cu
Inconel 604	...	16.0	74.0	...	...	...	2.25	...	...	7.5	0.02	0.03 max Cu
Inconel 617	...	22.0	55.0	12.5	9.0	...	...	...	1.0	...	0.07	...
Inconel 625	N06625	21.5	61.0	...	9.0	...	3.6	0.2	0.2	2.5	0.05	...
NA-224	...	27.0	48.0	...	...	6.0	...	...	...	18.5	0.50	...
Nimonic 75	...	19.5	75.0	...	...	...	...	0.4	0.15	2.5	0.12	0.25 max Cu
RA-333	N06333	25.0	45.0	3.0	3.0	3.0	...	...	...	18.0	0.05	...

(continued)

Table 3.2 (continued)

Alloy	UNS number	Cr	Ni	Co	Mo	W	Composition, % Nb	Ti	Al	Fe	C	Other
Iron-base precipitation-hardening alloys												
A-286	K66286	15.0	26.0	...	1.25	...	...	2.0	0.2	55.2	0.04	0.005 B; 0.3 V
Discaloy	K66220	14.0	26.0	...	3.0	...	...	1.7	0.25	55.0	0.06	...
Haynes 556	...	22.0	21.0	20.0	3.0	2.5	0.1	...	0.3	29.0	0.10	0.50 Ta; 0.02 La; 0.002 Zr
Incoloy 903	...	0.1 max	38.0	15.0	0.1	...	3.0	1.4	0.7	41.0	0.04	...
Pyromet CTX-1	...	0.1 max	37.7	16.0	0.1	...	3.0	1.7	1.0	39.0	0.03	...
V-57	...	14.8	27.0	...	1.25	...	...	3.0	0.25	48.6	0.08 max	0.01 B; 0.5 max V
W-545	K66545	13.5	26.0	...	1.5	...	...	2.85	0.2	55.8	0.08	0.05 B
Cobalt-base precipitation-hardening alloys												
AR-213	...	19.0	0.5 max	65.0	...	4.5	...	...	3.5	0.5 max	0.17	6.5 Ta; 0.15 Zr; 0.1 Y
MP-35N	R30035	20.0	35.0	35.0	10.0	...	...	...	...	...	...	...
MP-159	...	19.0	25.0	36.0	7.0	...	0.6	3.0	0.2	9.0	...	...
Nickel-base precipitation-hardening alloys												
Astroloy	...	15.0	56.5	15.0	5.25	...	...	3.5	4.4	<0.3	0.06	0.03 B; 0.06 Zr
D-979	N09979; K66979(a)	15.0	45.0	...	4.0	4.0	...	3.0	1.0	27.0	0.05	0.01 B
IN 100	N13100	10.0	60.0	15.0	3.0	...	...	4.7	5.5	<0.6	0.15	1.0 V; 0.06 Zr; 0.015 B
IN 102	N06102	15.0	67.0	...	2.9	3.0	2.9	0.5	0.5	7.0	0.06	0.005 B; 0.02 Mg; 0.03 Zr
Incoloy 901	N09901	12.5	42.5	...	6.0	...	...	2.7	...	36.2	0.10 max	...
Inconel 706	N09706	16.0	41.5	...	...	...	...	1.75	0.2	37.5	0.03	2.9 (Nb + Ta); 0.15 max Cu
Inconel 718	N07718	19.0	52.5	...	3.0	...	5.1	0.9	0.5	18.5	0.08 max	0.15 max Cu
Inconel 751	...	15.5	72.5	...	...	...	1.0	2.3	1.2	7.0	0.05	0.25 max Cu
Inconel X750	N07750	15.5	73.0	...	...	...	1.0	2.5	0.7	7.0	0.04	0.25 max Cu
M252	N07252	19.0	56.5	10.0	10.0	...	...	2.6	1.0	<0.75	0.15	0.005 B
Nimonic 80A	N07080	19.5	73.0	1.0	...	...	...	2.25	1.4	1.5	0.05	0.10 max Cu
Nimonic 90	N07090	19.5	55.5	18.0	...	...	...	2.4	1.4	1.5	0.06	...
Nimonic 95	...	19.5	53.5	18.0	...	...	...	2.9	2.0	5.0 max	0.15 max	+B; +Zr
Nimonic 100	...	11.0	56.0	20.0	5.0	...	...	1.5	5.0	2.0 max	0.30 max	+B; +Zr

(continued)

Table 3.2 (continued)

Alloy	UNS number	Cr	Ni	Co	Mo	W	Composition, % Nb	Ti	Al	Fe	C	Other
Nickel-base precipitation-hardening alloys												
Nimonic 105	...	15.0	54.0	20.0	5.0	...	...	1.2	4.7	...	0.08	0.005 B
Nimonic 115	...	15.0	55.0	15.0	4.0	...	...	4.0	5.0	1.0	0.20	0.04 Zr
Nimonic 263(b)	...	20.0	51.0	20.0	5.9	...	...	2.1	0.45	0.7 max	0.06	...
Pyromet 860	...	13.0	44.0	4.0	6.0	...	...	3.0	1.0	28.9	0.05	0.01 B
Refractory 26	...	18.0	38.0	20.0	3.2	...	...	2.6	0.2	16.0	0.03	0.015 B
René 41	N07041	19.0	55.0	11.0	10.0	...	...	3.1	1.5	<0.3	0.09	0.01 B
René 95	...	14.0	61.0	8.0	3.5	3.5	3.5	2.5	3.5	<0.3	0.16	0.01 B; 0.05 Zr
René 100	...	9.5	61.0	15.0	3.0	...	...	4.2	5.5	1.0 max	0.16	0.015 B; 0.06 Zr; 1.0 V
Udimet 500	N07500	19.0	48.0	19.0	4.0	...	...	3.0	3.0	4.0 max	0.08	0.005 B
Udimet 520	...	19.0	57.0	12.0	6.0	1.0	...	3.0	2.0	...	0.08	0.005 B
Udimet 630	...	17.0	50.0	...	3.0	3.0	6.5	1.0	0.7	18.0	0.04	0.004 B
Udimet 700	...	15.0	53.0	18.5	5.0	...	...	3.4	4.3	<1.0	0.07	0.03 B
Udimet 710	...	18.0	55.0	14.8	3.0	1.5	...	5.0	2.5	...	0.07	0.01 B
Unitemp AF2-1DA	...	12.0	59.0	10.0	3.0	6.0	...	3.0	4.6	<0.5	0.35	1.5 Ta; 0.015 B; 0.1 Zr
Waspaloy	N07001	19.5	57.0	13.5	4.3	...	...	3.0	1.4	2.0 max	0.07	0.006 B; 0.09 Zr

(a) No longer active, but shown here for reference. (b) Also known as Rolls Royce C-263.

Table 3.3 Compositions of nickel-base heat-resistant casting alloys

Alloy designation	Nominal composition, % C	Ni	Cr	Co	Mo	Fe	Al	B	Ti	W	Zr	Others
B-1900	0.1	64	8	10	6	...	6	0.015	1	...	0.10	4Ta(a)
Hastelloy X	0.1	50	21	1	9	18	...	...	...	1	...	...
IN-100	0.18	60.5	10	15	3	...	5.5	0.01	5	...	0.06	1V
IN-738X	0.17	61.5	16	8.5	1.75	...	3.4	0.01	3.4	2.6	0.1	1.75Ta, 0.9Nb
IN-792	0.2	60	13	9	2.0	...	3.2	0.02	4.2	4	0.1	4Ta
Inconel 713C	0.12	74	12.5	...	4.2	...	6	0.012	0.8	...	0.1	2Nb
Inconel 713LC	0.05	75	12	...	4.5	...	6	0.01	0.6	...	0.1	2Nb
Inconel 718	0.04	53	19	...	3	18	0.5	...	0.9	...	...	0.1Cu, 5Nb
Inconel X-750	0.04	73	15	...	...	7	0.7	...	2.5	...	...	0.25Cu, 0.9Nb
M-252	0.15	56	20	10	10	...	1	0.005	2.6	...	...	...
MAR-M 200	0.15	59	9	10	...	1	5	0.015	2	12.5	0.05	1Nb(b)
MAR-M 246	0.15	60	9	10	2.5	...	5.5	0.015	1.5	10	0.05	1.5Ta
MAR-M 247	0.15	59	8.25	10	0.7	0.5	5.5	0.015	1	10	0.05	1.5Hf, 3Ta
NX 188 (DS)	0.04	74	...	...	18	...	8	...	...	...	...	...
René 77	0.07	58	15	15	4.2	...	4.3	0.015	3.3	...	0.04	...
René 80	0.17	60	14	9.5	4	...	3	0.015	5	4	0.03	...
René 100	0.18	61	9.5	15	3	...	5.5	0.015	4.2	...	0.06	1V
TRW-NASA VIA	0.13	61	6	7.5	2	...	5.5	0.02	1	6	0.13	0.4Hf, 0.5Nb, 0.5Re, 9Ta
Udimet 500	0.1	53	18	17	4	2	3	...	3	...	...	...
Udimet 700	0.1	53.5	15	18.5	5.25	...	4.25	0.03	3.5	...	...	...
Udimet 710	0.13	55	18	15	3	...	2.5	...	5	1.5	0.08	...
Waspaloy	0.07	57.5	19.5	13.5	4.2	1	1.2	0.005	3	...	0.09	...
WAZ-20 (DS)	0.20	72	...	...	...	...	6.5	...	...	20	1.5	...

(a) B-1900 + Hf also contains 1.5% Hf. (b) MAR-M 200 + Hf also contains 1.5% Hf.

Table 3.4 Compositions of cobalt-base heat-resistant casting alloys

Alloy designation	Nominal composition, % C	Co	Cr	Ni	Al	B	Fe	Ta	W	Zr	Others
AiResist 13	0.45	62	21	...	3.4	...	...	2	11	...	0.1Y
AiResist 213	0.20	64	20	0.5	3.5	...	0.5	6.5	4.5	0.1	0.1Y
AiResist 215	0.35	63	19	0.5	4.3	...	0.5	7.5	4.5	0.1	0.1Y
Haynes 21	0.25	64	27	3	...	...	1	...	...	...	5 Mo
Haynes 25; L-605	0.1	54	20	10	...	...	1	...	15	...	...
Haynes 151(a)	0.48	65	20	...	...	0.03	...	...	12.8	...	3 max Fe + Ni
J-1650	0.20	36	19	27	...	0.02	...	2	12	...	3.8 Ti
MAR-M 302	0.85	58	21.5	...	...	0.005	0.5	9	10	0.2	...
MAR-M 322	1.0	60.5	21.5	...	...	...	0.5	4.5	9	2	0.75 Ti
MAR-M 509	0.6	54.5	23.5	10	...	...	...	3.5	7	0.5	0.2 Ti
MAR-M 918	0.05	52	20	20	...	...	...	7.5	...	0.1	...
NASA Co-W-Re	0.40	67.5	3	...	...	...	...	...	25	1	2 Re, 1 Ti
S-816	0.4	42	20	20	...	...	4	...	4	...	4 Mo, 4 Nb, 1.2 Mn, 0.4 Si
V-36	0.27	42	25	20	...	...	3	...	2	...	4 Mo, 2 Nb, 1 Mn, 0.4 Si
WI-52	0.45	63.5	21	...	...	...	2	...	11	...	2 Nb+Ta
X-40	0.50	57.5	22	10	...	...	1.5	...	7.5	...	0.5 Mn, 0.5 Si

(a) Obsolete alloy, included for reference purposes.

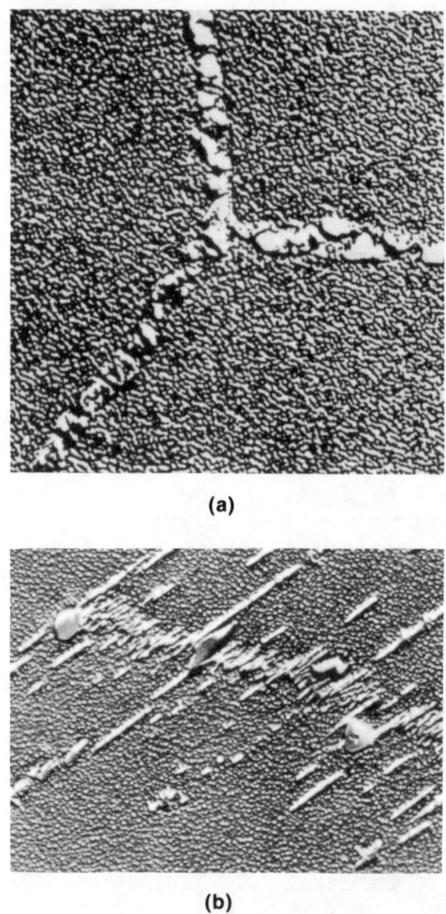

(a)

(b)

(a) 10,000×. (b) 6800×.

Fig. 3.1 Favorable discrete grain-boundary carbide precipitation (a) and less favorable zipperlike, discontinuous carbide precipitation (b) in Waspaloy specimens

fraction γ' ($\geq$0.35) nickel-base superalloys. The inherent strength capability of such superalloys is controlled by the intragranular distribution; however, the usable strength in polycrystalline alloys is determined by the condition of the grain boundaries, particularly as affected by the carbide-phase morphology and distribution. Satisfactory properties are achieved by optimizing the volume fraction and morphology of γ', which are not necessarily independent characteristics, in conjunction with securing a dispersion of discrete globular carbides along the grain boundaries (Fig. 3.1). Discontinuous (cellular) carbide or γ' at grain boundaries (Fig. 3.1) increases surface area and drastically reduces rupture life, even though tensile and creep strength may be relatively unaffected.

Wrought nickel and iron-nickel superalloys generally are processed to have optimum tensile and fatigue properties. At one time, when wrought alloys were used for creep-limited applications such as gas turbine high-pressure turbine blades, heat treatments that were different from those used for tensile-limited uses were applied to the same nominal alloy composition to maximize creep-rupture life. Occasionally, the nominal composition of an alloy such as IN-100 or Udimet 700 (Astroloy) varies according to whether it is to be used in the cast or the wrought condition.

EVOLUTION OF MICROSTRUCTURE

Superalloys contain a variety of elements in a large number of combinations to produce desired effects. Some elements go into solid solution to provide one or more of the following: strength (molybdenum, tantalum, tungsten, and rhenium); oxidation resistance (chromium and aluminum); phase stability (nickel); and increased volume fractions of favorable secondary precipitates (cobalt). Other elements are added to form strengthening precipitates such as γ' (aluminum or titanium) and γ'' (niobium). Minor elements (carbon or boron) are added to form carbides and borides; these and other elements (cerium or magnesium) are added for purposes of tramp-element control. Some elements (boron, zirconium, and hafnium) also are added to promote grain-boundary effects other than precipitation or carbide formation. Many elements (cobalt, molybdenum, tungsten, chromium, etc.), although added for their favorable alloying qualities, can participate, in some circumstances, in undesirable phase formation (σ, μ, Laves, etc.).

Some of these elements produce readily discernible changes in microstructure; other elements produce more subtle microstructural effects. The precise microstructural effects produced are functions of processing and heat treatment. The most obvious microstructural effects involve precipitation of geometrically close-packed (gcp) phases such as γ', formation of carbides, and formation of topologically close-packed (tcp) phases such as σ. Even when the type of phase is specified, microstructure morphology can vary widely; for example, script vs blocky carbides, cuboidal vs spheroidal γ', cellular vs uniform precipitation, acicular vs blocky σ, and discrete γ' vs γ' envelopes. Typical nickel-base superalloy microstructures as they evolved from spheroidal to cuboidal γ' are shown in Fig. 3.2.

The γ' phase is an intermetallic fcc phase having the basic composition $Ni_3(Al,Ti)$. Alloying elements affect γ' mismatch with the matrix γ phase, γ' morphology, and γ' stability. A related phase, eta (η), is a hexagonal phase of composition Ni_3Ti, which may exist in a metastable form as γ' before transforming to η. Other types of intermetallic phases such as delta (δ), orthorhombic Ni_3Cb or γ'', which is a body-centered tetragonal (bct) Ni_3Cb strengthening precipitate, have been observed.

Carbides also are an important constituent of superalloys. They are particularly essential in the grain boundaries of polycrystalline alloys for production of desired strength and ductility. Carbides also may provide some degree of matrix strengthening, particularly in cobalt-base alloys, and are necessary for grain-size control in some wrought alloys. Some carbides are virtually unaffected by heat treatment, whereas others require such treatment. Various types of carbides are possible depending on alloy composition and processing.

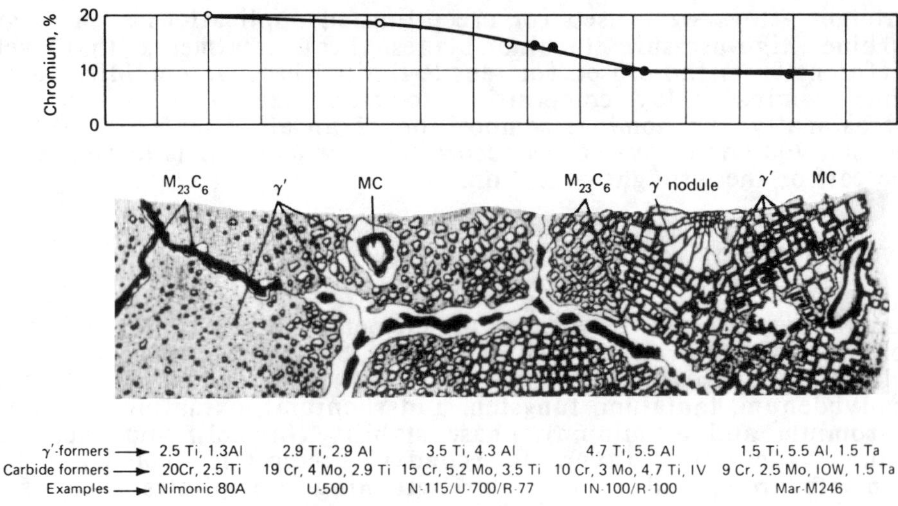

Adapted from *Heat Treatment, Structure and Properties of Nonferrous Alloys*, C.R. Brooks, American Society for Metals, 1982.

Fig. 3.2 Evolution of microstructure and chromium content of nickel-base superalloys

Some of the important types are MC, M_6C, $M_{23}C_6$, and M_7C_3, where "M" designates one or more types of metal atom. In many cases, the carbides exist jointly; however, they usually are formed by sequential reactions. The common carbide reaction sequence for many superalloys is MC to $M_{23}C_6$, and the important carbide-forming elements are chromium ($M_{23}C_6$ and M_7C_3), titanium, tantalum, niobium, and hafnium (MC), and molybdenum and tungsten (M_6C). Boron may participate somewhat interchangeably with carbon and produces such phases as MB_{12}, M_3B_2, and others. One claim made for boron is that primary borides formed by adjustment of the boron/carbon ratio are more amenable to morphological modification through heat treatment.

All superalloys contain some chromium plus other elements to promote resistance to environmental degradation. The role of chromium is to promote Cr_2O_3 formation on the external surface of an alloy. When sufficient aluminum is present, formation of the more protective oxide Al_2O_3 is promoted when oxidation occurs. A chromium content of 6 to 22 wt% generally is common in nickel-base superalloys, whereas a level of 20 to 30 wt% Cr is characteristic of cobalt-base superalloys. A level of 15 to 25 wt% Cr is found in iron-nickel superalloys. Amounts of aluminum up to 6 wt% may be present in nickel-base superalloys.

A discussion of the function of alloying elements on microstructure would be incomplete without mention of the tramp elements. Elements such as silicon, phosphorus, sulfur, lead, bismuth, tellurium, selenium, and silver, often in amounts as low as the parts-per-million (ppm) level, have been associated with property level reductions, but they are not visible optically or with an electron microscope. Microprobe and Auger spectroscopic analyses have determined that grain boundaries can be decorated with tramp elements at high local concentrations. Elements such as magnesium and cerium tend to tie up some detrimental elements such as sulfur in the form of a compound, and titanium tends to tie up nitrogen as TiN. In such cases, these and other similar compounds often are visible in the microstructure.

Function of Processing

Processing is considered to be the art/science of rendering the superalloy material into its final form. Processing and alloying elements are interdependent. The general microstructural changes brought about by processing result from the overall alloy composition plus the processing sequence. The role of heat treatment on phases will be referred to when prior microstructural effects are considered; the role of composition on phases has been discussed above.

The three most significant process-related microstructural variables other than those resulting from composition/heat treatment interactions are grain size, grain shape, and orientation (in anisotropic structures). Grain size varies considerably from cast to wrought structures, generally being significantly smaller for the latter. Special processing, e.g., directional solidification or directional recrystallization, can effect changes not only in grain size, but also in grain shape and orientation, which significantly alter mechanical and physical properties. Corrosion reactions, however, are primarily functions of composition.

EFFECTS OF PRIOR MICROSTRUCTURE ON PROPERTIES

There are five major effects on properties of superalloys that result from prior microstructure. These are:

* Gamma prime (γ') precipitation
* Carbide precipitation: grain-boundary strengthening
* Carbide precipitation: matrix or general strengthening
* Gamma double prime (γ'') precipitation
* Effects of boron, zirconium, and hafnium

Gamma Prime Precipitation

Nickel and iron-nickel superalloys may be strengthened by γ' precipitation (in an fcc γ matrix). The γ' in iron-nickel and first-generation nickel-base alloys generally is spheroidal, whereas the γ' in later generation nickel-base alloys generally is cuboidal. (See Fig. 3.2 and 3.3.) The V_f of γ' generally is about 0.2 or less in wrought iron-nickel superalloys, but may exceed 0.6 in nickel-base superalloys. Strengthening by γ' is related to many factors; the most direct correlations can be made with volume fraction of γ' and with γ' particle size. However, the correlation between strength and γ' size may be difficult to prove in commercial alloys over the range of particle sizes available.

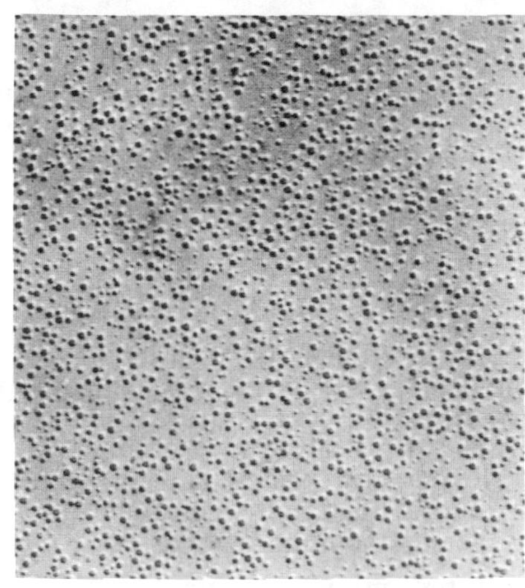

Note secondary (cooling) γ' between primary cuboidal γ' particles in Udimet 700. Original magnification, 6800×.

Fig. 3.3 Wrought nickel-base superalloys showing spheroidal nature of early (low volume fraction γ') alloys (Waspaloy, left) and cuboidal nature of later (higher volume fraction γ') alloys (Udimet 700, right)

Before the age-strengthening peak is reached during precipitation, the operative strengthening mechanism involves cutting of γ' particles by dislocations, and strength increases with γ' size (Fig. 3.4) at constant V_f of γ'. After the age-strengthening peak is reached, strength decreases with continuing particle growth, because dislocations no longer cut γ' particles but bypass them. This effect can be demonstrated for tensile or hardness behavior in low volume fraction γ' alloys (A-286, Inconel 901, and Waspaloy), but is not as readily apparent in high volume fraction γ' alloys, such as MAR-M 246, IN-100, etc. For creep-rupture, the effects are less well defined; however, uniform fine-to-moderate γ' sizes (0.25 to 0.5 μm) are preferred to coarse or hyperfine γ' for optimum properties.

Alloy strength clearly depends on the volume fraction of γ'. The volume fraction of γ' and thus strength can be increased to a point by adding more hardener elements (aluminum or titanium). Alloy strengths increase as Al + Ti content increases (Fig. 3.5) and also as the Al-Ti ratio increases. In wrought alloys, the γ' usually exists as a bimodal (duplex) distribution of fine γ', and all of the Al + Ti contributes effectively to the strengthening process. In cast alloys, the character of the γ' precipitate developed can be extremely variable because of the effects of segregation and cooling rate. Large amounts of γ-γ' eutectic and coarse γ' may be developed during solidification. Subsequent heat treatments can modify these structures. Bimodal and trimodal γ' distributions plus γ-γ' eutectic can be found in cast alloys after heat treatment. Solution heat treatments at temperatures sufficiently high to homogenize the alloy and dissolve coarse γ' and the eutectic γ-γ' constituents for subsequent reprecipitation as a uniform, fine γ' have improved creep capability. For a

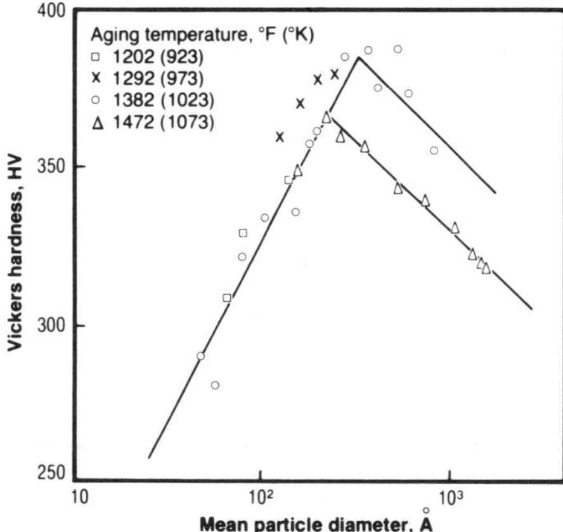

Cutting occurs at low particle sizes, bypassing at larger sizes. Note that aging temperature also affects strength in conjunction with particle size.

Fig. 3.4 Strength (hardness) vs particle diameter in a nickel-base superalloy

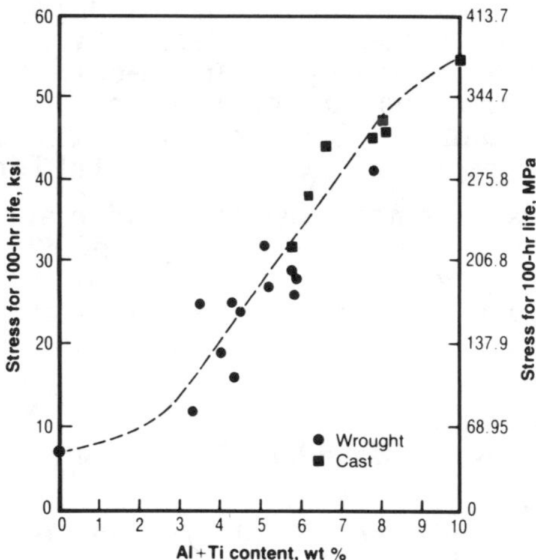

Fig. 3.5 Effect of Al + Ti content on strength of nickel-base superalloys at 870 °C (1600 °F)

columnar grain nickel-base superalloy, a direct correlation has been found to exist between creep-rupture life at 980 °C (1800 °F) and the V_f of fine γ' (Fig. 3.6).

In general, to achieve the greatest strengthening in γ'-strengthened alloys, it is necessary to solution heat treat the alloys above the γ' solvus. One or more aging treatments are employed to optimize the γ' distribution and to promote transitions in other phases such as carbides. In some alloys, several intermediate and several lower temperature aging treatments are used; in cast alloys or in very high

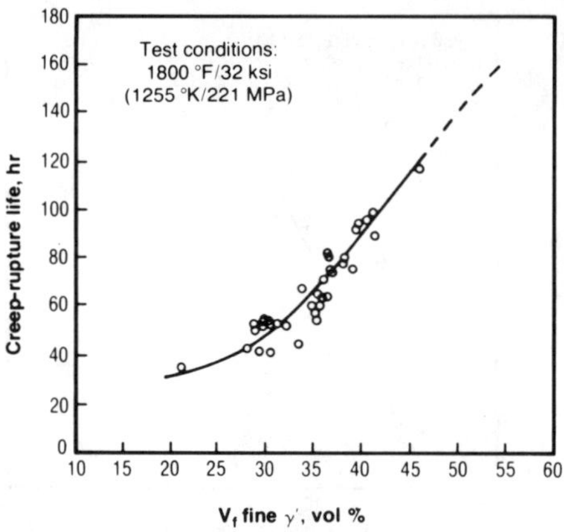

Fig. 3.6 Increase in creep-rupture life with increase in V_f of fine γ', demonstrated in a columnar grain, directionally solidified nickel-base superalloy

volume fraction γ' wrought alloys, a coating cycle or high-temperature aging treatment may precede the intermediate-temperature aging cycle. When multiple aging treatments are used, a superalloy may show the bimodal or trimodal γ' distribution described above. An essential feature of γ' hardening in nickel-base superalloys is that a temperature fluctuation that dissolves some γ' does not necessarily produce permanent damage, because subsequent cooling to normal operating conditions reprecipitates γ' in a useful form.

In the final analysis, it is not possible to judge alloy performance by considering only the γ' phase. The strength of γ'-strengthened grains must be balanced by grain-boundary strength. If the γ'-hardened matrix becomes much stronger relative to grain boundaries, then premature failure occurs because stress relaxation becomes difficult.

Carbide Precipitation/Grain-Boundary Strengthening

Carbides exert a profound influence on properties by their precipitation on grain boundaries. In most superalloys, $M_{23}C_6$ forms at the grain boundaries after a postcasting or postsolution treatment thermal cycle such as aging. A chain of discrete globular $M_{23}C_6$ carbides was found to optimize creep-rupture life by preventing grain-boundary sliding in creep-rupture while concurrently providing sufficient ductility in the surrounding grain for stress relaxation to occur without premature failure.

In contrast, if carbides precipitate as a continuous grain-boundary film, properties can be severely degraded. $M_{23}C_6$ films were reported to reduce impact resistance of M252, and MC films were blamed for lowered rupture lives and ductility in forged Waspaloy. At the other extreme, when no grain-boundary carbide precipitate is present, premature failure also will occur, because grain-boundary movement essentially is unrestricted, leading to subsequent cracking at grain-boundary triple points.

The role of carbides at grain boundaries in iron-nickel superalloys is less well documented than for nickel superalloys, although detrimental effects of carbide films have been reported. Studies of specific effects of grain-boundary carbides in cobalt-base alloys are even more sparse, because the carbide distribution in cobalt-base alloys arises from the original casting or on cooling after mill annealing for wrought cobalt-base alloys. The significantly greater carbon content of cobalt-base alloys leads to much more extensive grain-boundary carbide precipitation than in nickel and iron-nickel alloys. Carbides at grain boundaries in cast cobalt-base alloys appear as eutectic aggregates of M_6C, $M_{23}C_6$ and fcc γ-cobalt-base solid solution. No definitive study of the effects of varied carbide forms in grain boundaries on the mechanical behavior of cobalt-base superalloys has been reported.

The lamellar eutectic (carbides - γCo) nature of carbides ($M_{23}C_6$-M_6C) in cast cobalt-base superalloys is interesting. A somewhat similar morphology of $M_{23}C_6$, occurring when it is precipitated in cellular form in nickel and iron-nickel alloys, leads to mechanical property loss in such alloys. However, lamellar eutectic does not seem to degrade cast cobalt-base alloy properties. Cellular growth in nickel-base alloys was found to occur when a high supersaturation of carbon produced by solution heat treatment was not relieved by an intermediate precipitation treatment prior to aging at 705 to 760 °C (1300 to 1400 °F). The ductility of a nickel-base superalloy also was impaired by a different type of precipitation, namely Widmanstätten M_6C at grain and twin boundaries. However, formation of intragranular Widmanstätten M_6C after exposure of B-1900 nickel-base alloy did not appear to reduce properties.

Another effect produced by grain-boundary $M_{23}C_6$ carbide precipitation is the occasional formation, on either side of the boundary, of a zone depleted in γ' precipitate. These precipitate-free zones may have significant effects on rupture life of nickel and iron-nickel superalloys. If such zones should become wide or much weaker than the matrix, deformation would concentrate there, resulting in early failure. The more complex (higher volume fraction γ') alloys do not exhibit significant precipitate-free zone effects, probably because of their higher saturation with regard to γ'-forming elements. An effect seen concurrently with the precipitate-free zone and not clearly separated from it is the γ' envelope produced by breakdown of TiC and consequent formation of $M_{23}C_6$ or M_6C + γ' (from the excess titanium). Not only is the role of the γ' envelope insufficiently established, but there is also the remote possibility that the excess Ti-rich area is really either η or a metastable γ' which could transform to η in use.

Carbide Precipitation/Matrix or General Strengthening

Carbides affect the creep-rupture strengths of cobalt-base superalloys and some nickel and iron-nickel superalloys by formation within grains. In cobalt-base cast superalloys, script MC carbides are liberally interspersed within grains, causing a form of dispersion hardening that is not of a large magnitude owing to its relative coarseness. Distribution of carbides in cast alloys can be modified by heat treatment, but strength levels attained at all but the highest

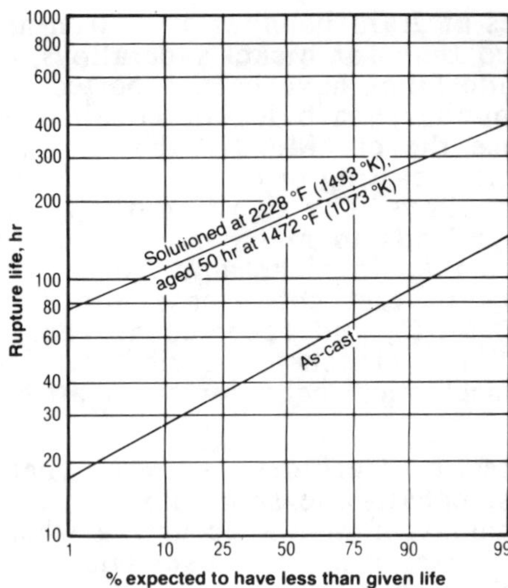

Fig. 3.7 Effect of heat treatment on a cobalt-base superalloy (HS-31), showing increase in strength resulting from carbide precipitation

temperatures are substantially less than those of the γ'-hardened alloys. Consequently, cast cobalt-base alloys generally are not heat treated, except in a secondary sense through the coating diffusion heat treatment of 4 h at 1065 to 1120 °C (1950 to 2050 °F) that is sometimes applied.

Wrought cobalt-base superalloys contain carbide modifications that are produced during fabrication. Carbide distributions in wrought alloys result from the mill anneal after final working. Properties are largely a result of grain size, refractory metal content, and carbon level, which indicates the V_f of carbides available for strengthening.

True solutioning, in which all minor constituents are dissolved, is not possible in most cobalt-base superalloys, because melting often occurs before all the carbides are solutioned. Some enhancement of creep-rupture behavior has been achieved by heat treatment. Rupture-time improvements can be gained by aging a modified Vitallium alloy (Fig. 3.7); larger increases have been produced by increasing the carbon content of the alloy. Solution treating and aging is not suitable for producing stable cobalt-base superalloys for use above 815 °C (1500 °F) because of carbide dissolution or over-aging.

Matrix carbides in nickel and iron-nickel superalloys also may be partially solutioned. MC carbides will not totally dissolve, however, without incipient melting of the alloy. Furthermore, MC carbides in such alloys tend to be unstable, because they decompose to $M_{23}C_6$ at temperatures below about 815 to 870 °C (1500 to 1600 °F), or possibly convert to M_6C at temperatures of 980 to 1040 °C (1800 to 1900 °F) if the alloy has a sufficiently high molybdenum-plus-tungsten content. Matrix carbides generally contribute very small increments of strengthening to nickel and iron-nickel superalloys.

An interesting microstructural trend has taken place with the advent of single crystals of nickel-base superalloys. Because no grain boundaries exist, there is little need for the normal grain-boundary strengtheners such as carbon. Consequently, very few matrix or sub-boundary carbides exist in such alloys.

Perhaps the most common other role of matrix carbides (also shared by grain-boundary carbides) is a negative one: they may participate in the fatigue-cracking process by premature cracking, or by oxidizing at the surface of uncoated alloys to cause a notch effect. Oxidized carbides or precracked carbides from machining or thermal stresses can initiate fatigue cracks. Precracked carbides can be related to prior casting processes. Carbide size is important, and reduced carbide volumes and sizes in nickel-base alloys result in a reduction in precracked carbides. The longer solidification times and lower gradients of early directional solidification processes often resulted in moderately large carbides. However, improved gradients and the reduced carbon contents of single-crystal alloys (few or no carbides) have resulted in substantial improvements in fatigue resistance - particularly over similarly oriented columnar grain alloys with normal carbon levels. This effect is most noticeable in low-cycle fatigue (LCF) and thermomechanical fatigue (TMF). No evidence is available to interpret the effect of the absence of carbides on high-cycle fatigue (HCF), but beneficial effects could be anticipated. Oxidized carbides can be minimized or prevented by several methods. Casting procedures and/or chemical composition may be modified to produce smaller primary carbides. Powder metallurgy (P/M) processing may be used to produce the same result. Carbon content may be reduced if it is not specifically required to enable the alloy to attain the desired strength levels. Reduced carbon is the rule in single-crystal and powdered superalloys. Of course, the alloy may be coated with an appropriate protective coating that leaves the carbides in a subsurface location.

Although there is limited documentation, it frequently is assumed that non-carbide-forming elements do influence the formation of carbides. Cobalt, for example, has been claimed to modify the carbides in nickel-base alloys, and phosphorus has produced a more general, more finely dispersed and smaller carbide precipitation than carbon alone in a heat-resistant iron-base alloy. The modifying effect on carbides may be intragranular or intergranular, depending on the modifier and the base-alloy system.

Gamma Double Prime Precipitation

The practical use of γ'' precipitation is restricted to iron-nickel superalloys with niobium additions. Inconel 718 is the outstanding example of an alloy in which γ'' formation has been exploited commercially. The γ'' phase is disk shaped, and the V_f of γ'' in Inconel 718 is substantially in excess of that of γ'. Both γ'' and γ' phases are found in alloys where γ'' is present, but γ'' can be the predominant strengthening agent. Although the strength of γ'' phase has not been studied, similar considerations to γ' behavior probably pertain. The most significant feature of γ'' is probably the ease with which it forms at moderate temperatures after prior solutioning by heat treatment or joining processes. Because of this behavior, an

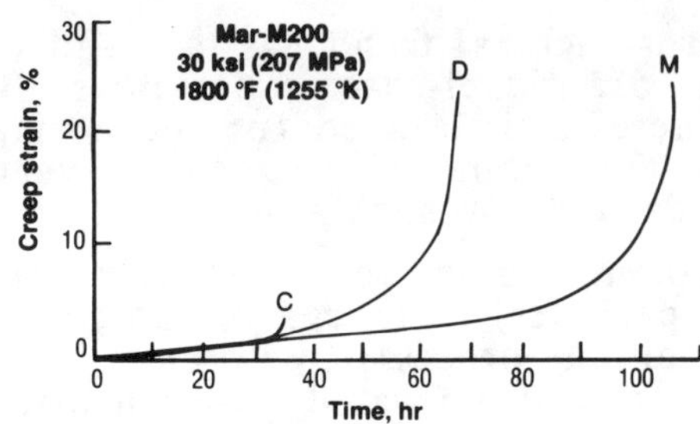

Fig. 3.8 Comparison of creep properties of polycrystalline conventionally cast (C), columnar grain directionally solidified (D), and single-crystal directionally solidified (M) MAR-M 200 alloy

alloy can be aged, after welding, to produce a fully strengthened structure with exceptional ductility.

The γ'' phase, not normally a stable phase, can convert to γ' or to δ Ni_3Cb on long-time exposure. The strength of γ'' is additive to that of γ' phase. A lack of notch ductility in Inconel 718 has been associated with a γ'' precipitate-free zone. The γ'' precipitate-free zone can be eliminated, and ductility can be restored by appropriate heat treatment. Alloys hardened with γ'' phase achieve high tensile strengths and very good creep-rupture properties at lower temperatures, but the conversion of γ'' to γ' or δ above temperatures of about 675 °C (1250 °F) causes a sharp reduction in strength.

Effects of Boron, Zirconium, and Hafnium

Within limits, significant improvements in mechanical properties can be achieved by additions of boron, zirconium, and hafnium. However, only limited microstructural correlations can be made. The presence of these elements may modify the initial grain-boundary carbides or tie up deleterious elements such as sulfur and lead. Reduced grain-boundary diffusion rates may be obtained, with consequent suppression of carbide agglomeration and creep cracking. Hafnium contributes to the formation of more γ-γ' eutectic in cast alloys. The eutectic at grain boundaries is thought (in modest quantities) to contribute to alloy ductility. The effects of these elements are limited to nickel and iron-nickel superalloys; virtually no cobalt-base alloys contain them.

EFFECTS OF PROCESSING

Three major processing techniques are used for controlling superalloy properties. Thermomechanical working is used for wrought nickel and iron-nickel alloys to store energy by producing a fine grain size and by controlling dislocation density/configuration. Improvements in tensile properties and low-cycle fatigue have resulted. A second processing route involves the use of P/M technology to produce reduced carbide size and greater homogeneity in materials, with a resulting improvement in fatigue resistance. Furthermore, in conjunction with superplastic forming, alloy grain sizes of ASTM 8 to 12 are being produced routinely in very high strength alloys, resulting in additional fatigue life benefits. Major benefits result from the ability to form alloys such as IN-100, which is unforgeable by some standard procedures.

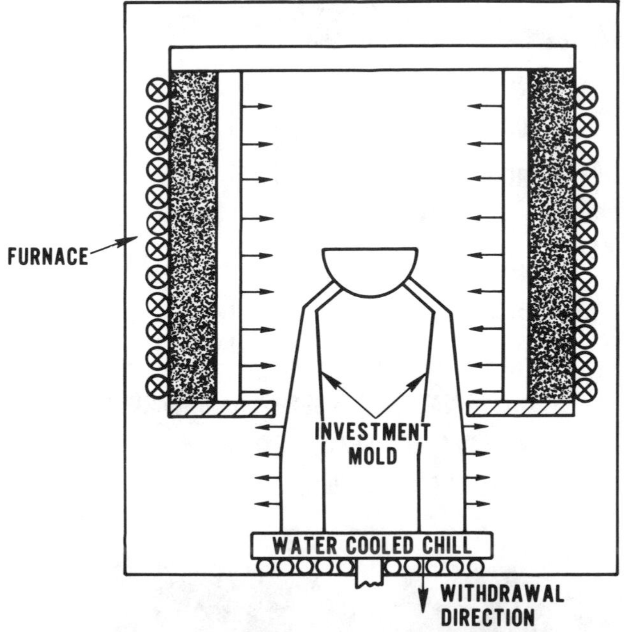

PRODUCTION D.S. PROCESS

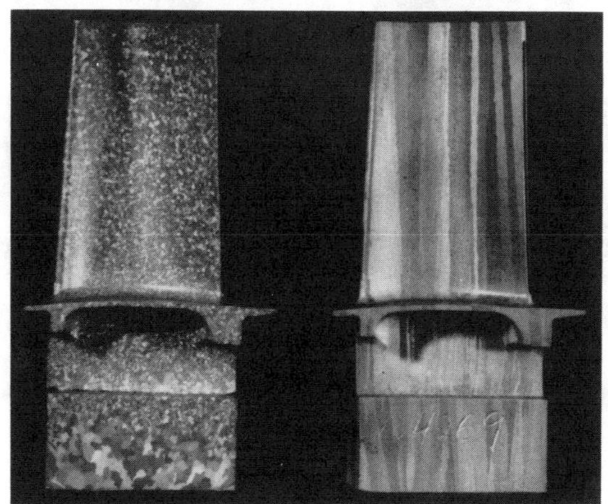

CONVENTIONAL AND D.S. BLADE

Courtesy of Pratt & Whitney Aircraft.

Fig. 3.9 Schematic of directional solidification process and resulting improvement in surface properties vs a conventional casting

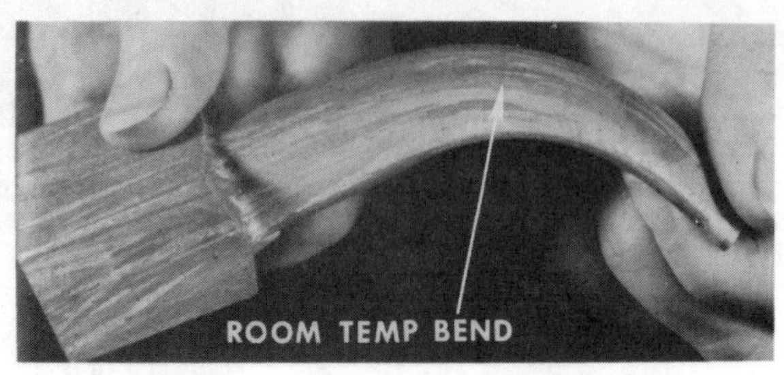

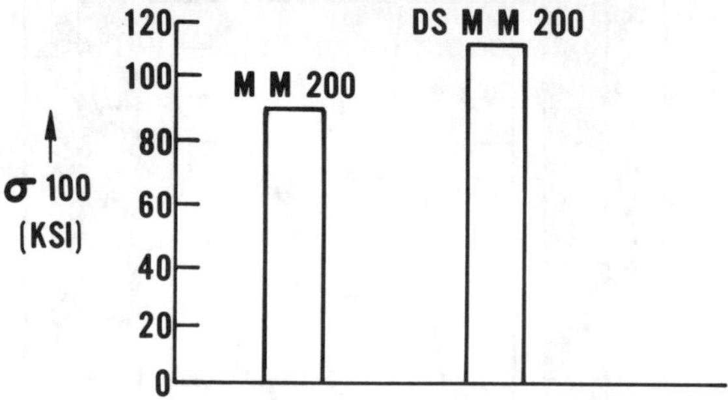

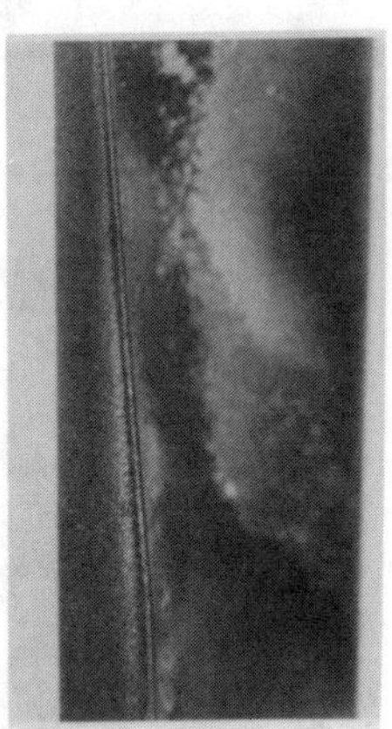

MM 200

DS MM 200

600 ENGINE CYCLES

● MAXIMIZED THERMAL FATIGUE RESISTANCE

Courtesy of Pratt & Whitney Aircraft.

Fig. 3.10 Effect of directional solidification on the physical properties of MAR-M 200

The third area is casting control of grain size and morphology, especially by controlled solidification. Cast alloy grain sizes have been made more uniform and, in some cases, have been reduced to enhance fatigue or tensile properties. Directional grain structures have improved strength (Fig. 3.8). Improved creep-rupture and fatigue resistance have resulted from the elimination of transverse grain boundaries and the favorable orientation of a low modulus direction to reduce strains. In the extreme, grains have been eliminated in single crystals with additional increases in creep-rupture behavior (Fig. 3.8). Figure 3.9 illustrates the production directional solidification process, and Fig. 3.10 illustrates the specific benefits realized in the production of MAR-M 200 castings.

Porosity in superalloys has led to fatigue and creep-rupture failure. Reduced porosity, achieved by hot isostatic pressing (HIP), also has resulted in improved properties. Efforts to date have centered on nickel-base cast alloys, but the process should enhance properties of any cast alloy with nonsurface-connected casting porosity. In the biomedical field, use of HIP has provided significant improvements in fatigue life of cast Vitallium alloy hip replacements.

Chapter 4

Property Overview

MATERIALS SELECTION

To satisfy the requirements of the wide variety of uses for which superalloys have been employed, extensive property data have been generated - not only for the typical mechanical, physical, chemical, nuclear, and magnetic properties, but also for several uncommon properties. The more important physical properties of these alloys include:

* Density
* Liquidus temperature
* Solidus temperature
* Specific heat
* Electrical conductivity
* Electrical resistivity
* Magnetic permeability
* Curie temperature

Superalloy density is influenced, of course, by alloying additions. Higher amounts of the light elements such as aluminum, titanium, and chromium reduce density, whereas the heavy solid-solution strengthening elements tungsten and tantalum increase density. Table 4.1 summarizes the physical properties of select superalloys.

Mechanical properties of importance (at specific temperatures) include:

* Tensile strength
* Yield strength
* Elongation
* Creep strength
* Stress-rupture strength
* Fatigue strength

Table 4.1 Physical properties of select superalloys

Alloy	Density, Mg/m³	Melting temperatures: Liquidus °C	Liquidus °F	Solidus °C	Solidus °F	Specific heat (a) J/kg	Specific heat (a) Btu/lb	Electrical conductivity, % IACS	Electric resistivity, nΩ·m	Magnetic permeability	Curie temperature °C	Curie temperature °F
Iron-base alloys												
Haynes 556	8.23	...	...	...	...	472	0.113	...	970	...	...	...
Incoloy 800	7.94	1385	2525	1355	2475	502	0.117	1.7	989	1.0092	...	...
Incoloy 801	7.94	1385	2525	1355	2475	452	0.105	1.7	1012	...	...	...
Incoloy 825	8.14	1400	2500	1370	2500	...	...	1.5	1127	1.005	< −196	< −520
Cobalt-base alloys												
Haynes 25 (L-605)	9.13	1410	2570	1329	2425	374	0.090	...	890	<1.00	...	...
Haynes 188	9.13	1398	2550	1302-1330	2375-2425	423(b)	0.101(b)	...	...	1.01	...	...
Stellite 6B	8.38	1354	2470	1265	2310	421	0.101	...	910	<1.2	...	...
UMCo 50	8.05	1395	2540	1380	2515			...	825	...	...	...
Nickel-base alloys												
Hastelloy B-2	9.21	...	...	...	...	389(b)	0.093(b)	...	1380(b)	...	...	...
Hastelloy C-4	8.64	...	...	...	...	426(b)	0.102(b)	...	1250	...	...	...
Hastelloy C-276	8.90	1371	2500	1323	2415	427	0.102	...	1330	...	...	...
Hastelloy N	8.93	...	...	...	...	419(b)	0.100(b)	...	1200(b)	...	...	...
Hastelloy S	8.76	1380	2516	1335	2435	427(b)	0.102(b)	...	...	...	...	...
Hastelloy W	9.03	1315	2400	...	...	...	...	...	...	...	...	...
Hastelloy X	8.23	1290	2350	1250	2280	486	0.116	...	1180	<1.002(c)	...	...
Inconel 600	8.42	1415	2575	1354	2470	444	0.103	1.7	1030	1.010	−124	−192
Inconel 617	...	...	...	1333	2430	...	...	...	...	...	...	...
Inconel 625	8.44	1350	2460	1290	2350	410	0.095	1.3	1290	1.006	−196	−320
Inconel 671	7.86	1350	2460	1305	2385	456	0.106	2.0	869	...	...	...
Inconel 690	8.03	1375	2510	1345	2450	...	...	1.5	148	...	...	...
Inconel X750	8.25	1425	2600	1393	2540	431	0.103	...	1215	1.0020	−143	−225
Nimonic 75	...	...	...	1380	2515	...	...	...	...	...	...	...
Nimonic 80A	...	...	...	1360	2480	...	...	...	...	...	...	...
Nimonic 90	...	...	...	1310	2390	...	...	...	...	...	...	...
Nimonic 100	...	...	...	...	...	...	...	...	...	...	...	...
Nimonic 105	...	...	...	1290	2256	...	...	...	...	...	...	...
René 41	8.25	1371	2500	1232	2250	452	0.108	...	1308	1.002	...	...
Udimet 500	8.14	1345	2450	1260	2300	...	...	...	1203	...	...	...
Udimet 700	7.92	1345	2450	1216	2220	...	...	...	...	...	...	...
Waspaloy	8.20	1355	2475	1339	2425	523(c)	0.125(c)	...	1240	...	...	...

(a) At room temperature. (b) At 100 °C (212 °F). (c) At 93 °C (200 °F).

TENSILE PROPERTIES

Table 4.2 compares the tensile properties of select superalloys, and Table 4.3 compares rupture strengths. Stress-rupture properties for the various superalloys also are presented in Fig. 4.1 to 4.5.

Short-time tensile and stress-rupture properties for several different product forms of Haynes 25 are compared in Table 4.4. Some alloys are more sensitive than others to differences in product form. Regardless of the alloy, the difference in properties caused by product form is likely to be more significant for stress-rupture properties than for short-time tensile properties.

The data in Table 4.4 are from different heats of metal and are affected by the different fabrication procedures used to produce the products, as well as by the shape of the test specimens, but they serve to illustrate the general relationship among product forms for a typical oxidation-resistant alloy.

Heat-resistant alloys are cold worked by hammering, swaging, and rolling. Cold worked products such as bolts and studs usually are stress relieved but, when tested at elevated temperature, retain a portion of the additional tensile and yield strengths imparted by the cold reduction.

Short-time tensile properties for the cobalt-base alloy Haynes 25 in the form of sheet 1.4 mm (0.054 in.) thick, solution treated at 1220 °C (2225 °F), showed a gradual and significant increase in strength as the amount of subsequent cold reduction was increased. The rupture properties were not improved (Table 4.5).

CREEP/STRESS-RUPTURE

Because superalloys are used primarily at elevated temperatures, high-temperature strength behavior is important. (Cryogenic properties and applications for superalloys are covered in Chapter 5.) In service at elevated temperature, the life of a metal component that is subjected to either static or dynamic loading is predictably limited. In contrast, at lower temperatures and in the absence of a corrosive environment, the life of a part is unlimited under static conditions, provided the operational loads do not exceed the yield strength of the metal. Stress imposed at elevated temperature, however, produces a continuous strain in the component and results in creep. By definition, creep is time-dependent strain or deformation occurring under stress at elevated temperature. After a period of time, creep terminates in fracture called stress-rupture.

The conditions of temperature, stress, and time under which creep and stress-rupture failures occur depend on the metal and the service environment. Consequently, elevated temperature failures may occur over a wide range of temperatures. In general, creep occurs at a temperature slightly above the recrystallization temperature of the metal involved. Elevated temperature behavior - the mechanical strength of the metal becoming limited by creep rather than by yield strength - must be determined for each material on the basis of individual characteristics. Parts can fail at elevated temperature for many reasons other than creep and stress rupture, such as low-cycle

Table 4.2 Typical tensile properties of select superalloys

Temperature		Tensile strength		Yield strength		Elongation,
°C	°F	MPa	ksi	MPa	ksi	%
		Cobalt-base alloys				
Haynes 25 (L-605) sheet						
21	70	1010	146	460	67	64
540	1000	800	116	250	36	59
650	1200	710	103	240	35	35
760	1400	455	66	260	38	12
870	1600	325	47	240	35	30
Haynes 188, sheet						
21	70	960	139	485	70	56
540	1000	740	107	305	44	70
650	1200	710	103	305	44	61
760	1400	635	92	290	42	43
870	1600	420	61	260	38	73
S-816, bar						
21	70	965	140	385	56	30
540	1000	840	122	310	45	27
650	1200	765	111	305	44	25
760	1400	650	94	285	41	21
870	1600	360	52	240	35	16
		Nickel-base alloys				
Astroloy, bar						
21	70	1410	205	1050	152	16
540	1000	1240	180	965	140	16
650	1200	1310	190	965	140	18
760	1400	1160	168	910	132	21
870	1600	770	112	690	100	25
D-979, bar						
21	70	1410	204	1010	146	15
540	1000	1300	188	925	134	15
650	1200	1100	160	890	129	21
760	1400	7	104	655	95	17
870	1600	345	50	305	44	18
Hastelloy X, sheet						
21	70	785	114	360	52	43
540	1000	650	94	290	42	45
650	1200	570	83	275	40	37
760	1400	435	63	260	38	37
870	1600	255	37	180	26	50
IN-102, bar						
21	70	960	139	505	73	47
540	1000	825	120	400	58	48
650	1200	710	103	400	58	64
760	1400	440	64	385	56	110
870	1600	215	31	200	29	110

(continued)

Table 4.2 (continued)

Temperature °C	Temperature °F	Tensile strength MPa	Tensile strength ksi	Yield strength MPa	Yield strength ksi	Elongation, %
Inconel 600, bar						
21	70	620	90	250	36	47
540	1000	580	84	195	28	47
650	1200	450	65	180	26	39
760	1400	185	27	115	17	46
870	1600	105	15	62	9	80
Inconel 601, sheet						
21	70	740	107	340	49	45
540	1000	725	105	150	22	38
650	1200	525	76	180	26	45
760	1400	290	42	200	29	73
870	1600	160	23	140	20	92
Inconel 625, bar						
21	70	855	124	490	71	50
540	1000	745	108	405	59	50
650	1200	710	103	420	61	35
760	1400	505	73	420	61	42
870	1600	285	41	475	40	125
Inconel 706, bar						
21	70	1300	188	980	142	19
540	1000	1120	163	895	130	19
650	1200	1010	147	825	120	21
760	1400	690	100	675	98	32
Inconel 718, bar						
21	70	1430	208	1190	172	21
540	1000	1280	185	1060	154	18
650	1200	1230	178	1020	148	19
760	1400	950	138	740	107	25
870	1600	340	49	330	48	88
Inconel 718, sheet						
21	70	1280	185	1050	153	22
540	1000	1140	166	945	137	26
650	1200	1030	150	870	126	15
760	1400	675	98	625	91	8
Inconel X 750, bar						
21	70	1120	162	635	92	24
540	1000	965	140	580	84	22
650	1200	825	120	565	82	9
760	1400	485	70	455	66	9
870	1600	235	34	165	24	47
M-252, bar						
21	70	1240	180	840	122	16
540	1000	1230	178	765	111	15
650	1200	1160	168	745	108	11
760	1400	945	137	715	104	10
870	1600	510	74	485	70	18

(continued)

Table 4.2 (continued)

Temperature °C	°F	Tensile strength MPa	ksi	Yield strength MPa	ksi	Elongation, %
Nimonic 75, bar						
21	70	750	109	...	...	41
540	1000	635	92	...	...	41
650	1200	538	78	...	...	42
760	1400	290	42	...	...	70
870	1600	145	21	...	...	68
Nimonic 80A, bar						
21	70	1240	179	620	90	24
540	1000	1100	160	530	77	24
650	1200	1000	145	550	80	18
760	1400	760	110	505	73	20
870	1600	400	58	260	38	34
Nimonic 90, bar						
21	70	1240	180	805	117	23
540	1000	1100	160	725	105	23
650	1200	1030	150	685	99	20
760	1400	825	120	540	78	10
870	1600	430	62	260	38	16
Nimonic 105, bar						
21	70	1140	166	815	118	12
540	1000	1100	160	775	112	18
650	1200	1080	156	800	116	24
760	1400	965	140	655	95	22
870	1600	605	88	365	53	25
Nimonic 115, bar						
21	70	1240	180	860	125	25
540	1000	1090	158	795	115	26
650	1200	1120	163	815	118	25
760	1400	1080	157	800	116	22
870	1600	825	120	550	80	18
Pyromet 860, bar						
21	70	1300	188	835	121	22
540	1000	1250	182	840	122	15
650	1200	1110	161	850	123	17
760	1400	910	132	835	121	18
René 41, bar						
21	70	1420	206	1060	154	14
540	1000	1400	203	1010	147	14
650	1200	1340	194	1000	145	14
760	1400	1100	160	940	136	11
870	1600	620	90	550	80	19
René 95, bar						
21	70	1620	235	1310	190	15
540	1000	1540	224	1250	182	12
650	1200	1460	212	1220	177	14
760	1400	1170	170	1100	160	15

(continued)

Table 4.2 (continued)

Temperature °C	°F	Tensile strength MPa	ksi	Yield strength MPa	ksi	Elongation, %
Udimet 500, bar						
21	70	1310	190	840	122	32
540	1000	1240	180	795	115	28
650	1200	1210	176	760	110	28
760	1400	1040	151	730	106	39
870	1600	640	93	495	72	20
Udimet 520, bar						
21	70	1310	190	860	125	21
540	1000	1240	180	825	120	20
650	1200	1170	170	795	115	17
760	1400	725	105	725	105	15
870	1600	515	75	515	75	20
Udimet 700, bar						
21	70	1410	204	965	140	17
540	1000	1280	185	895	130	16
650	1200	1240	180	855	124	16
760	1400	1030	150	825	120	20
870	1600	690	100	635	92	27
Udimet 710, bar						
21	70	1190	172	910	132	7
540	1000	1150	167	850	123	10
650	1200	1290	187	860	125	15
760	1400	1020	148	815	118	25
870	1600	705	102	635	92	29
Unitemp AF2-1DA, bar						
21	70	1290	187	1050	152	10
540	1000	1340	194	1080	157	13
650	1200	1360	197	1080	157	13
760	1400	1150	167	1010	146	8
870	1600	830	120	715	104	8
Waspaloy, bar						
21	70	1280	185	795	115	25
540	1000	1170	170	725	105	23
650	1200	1120	162	690	100	34
760	1400	795	115	675	98	28
870	1600	525	76	515	75	35

and high-cycle fatigue, thermal fatigue, tension overload, and combinations of these conditions. However, it is the creep/stress-rupture behavior that differentiates between low- and high-temperature conditions.

Obviously then, most parts operating at elevated temperature cannot be designed on the basis of tensile properties, but rather they must be designed using creep and stress-rupture data. High-temperature tensile properties are valuable for a quick comparison of superalloy strengths, for use when operating temperatures are low enough that creep and stress-rupture behavior is not involved, or when short-time

Table 4.3 Typical rupture strengths of select superalloys

Temperature		For stress rupture at: 100 h		1000 h	
°C	°F	MPa	ksi	MPa	ksi
Incoloy 800					
650	1200	220	32	145	21
760	1400	115	17	69	10
870	1600	45	6.5	33	4.8
Incoloy 801					
650	1200	250	36	...	...
730	1350	145	21	...	...
815	1500	62	9	...	...
Incoloy 802					
650	1200	240	35	170	24
760	1400	145	21	105	15
870	1600	97	14	62	9
Inconel 600					
815	1500	55	8	39	5.6
870	1600	37	5.3	24	3.5
Inconel 601(a)					
540	1000	...	...	400	58
870	1600	48	7	30	4.3
980	1800	23	3.4	14	2.1
Inconel 617(b)					
815	1500	140	20	97	14
925	1700	62	9	...	5.5
980	1800	41	6	...	3.5
Inconel 625(a)					
650	1200	440	64	370	54
815	1500	130	19	93	13.5
870	1600	72	10.5	48	7
Inconel 718(c)					
540	1000	...	...	951	138
595	1100	860	125	760	110
650	1200	690	100	585	85
Inconel 751(d)					
815	1500	200	29	125	185
870	1600	120	175	69	10
Inconel X750(e)					
540	1000	...	...	827	120
870	1600	83	12	45	6.5
925	1700	58	8.4	21	3.1

(continued)

(a) Solution treated 1150 °C (2100 °F). (b) Solution treated 1175 °C (2150 °F). (c) Heat treated to 980 °C (1800 °F) plus 720 °C (1325 °F) hold for 8 h, F.C. to 620 °C (1150 °F) hold for 8 h. (d) 730 °C (1350 °F) hold for 2 h. (e) Heat treat to 1150 °C (2100 °F) plus 840 °C (1550 °F) hold for 24 h, plus 705 °C (1300 °F) hold for 20 h. (f) Solution treated and aged. (g) Stress-relieved forging. (h) Heat treat to 1050 °C (1922 °F) hold for 1 h. (j) Heat treat to 1080 °C (1976 °F) hold for 8 h, plus 700 °C (1290 °F) hold for 16 h. (k) Heat treat to 1150 °C (2100 °F) hold for 4 h, plus 1050 °C (1920 °F) hold for 16 h, plus 850 °C (1560 °F) hold for 16 h. (m) Heat treat to 1190 °C (2175 °F) hold for 1.5 h, plus 1100 °C (2010 °F) hold for 6 h. (n) Heat treat to 1150 °C (2100 °F) hold for 2 h, W.Q., plus 800 °C (1475 °F) hold for 8 h.

Table 4.3 (continued)

Temperature °C	Temperature °F	For stress rupture at: 100 h MPa	100 h ksi	1000 h MPa	1000 h ksi
N-155, bar(f)					
650	1200	360	52	295	43
730	1350	195	28	150	22
870	1600	97	14	66	9.5
N-155(g)					
650	1200	380	55	290	42
N-155, sheet(f)					
980	1800	39	5.6	20	2.9
Nimonic 75(h)					
815	1500	38	5.5	24	3.5
870	1600	23	3.4	15	2.2
925	1700	14	2.1	10	1.5
980	1800	...	...	7.6	1.1
Nimonic 80A(j)					
540	1000	...	...	825	120
815	1500	185	27	115	17
870	1600	105	15	...	...
Nimonic 90(j)					
815	1500	240	35	155	22.5
870	1600	150	22	69	10
925	1700	69	10	...	...
Nimonic 105(k)					
815	1500	325	47	225	32
870	1600	210	30.2	135	19
Nimonic 115(m)					
815	1500	425	62	315	46
870	1600	315	46	205	30
925	1700	205	30	130	18.5
Nimonic 263(n)					
815	1500	170	24.5	105	15
870	1600	93	13.5	46	6.7
925	1700	45	6.5	...	...

(a) Solution treated 1150 °C (2100 °F). (b) Solution treated 1175 °C (2150 °F). (c) Heat treated to 980 °C (1800 °F) plus 720 °C (1325 °F) hold for 8 h, F.C. to 620 °C (1150 °F) hold for 8 h. (d) 730 °C (1350 °F) hold for 2 h. (e) Heat treat to 1150 °C (2100 °F) plus 840 °C (1550 °F) hold for 24 h, plus 705 °C (1300 °F) hold for 20 h. (f) Solution treated and aged. (g) Stress-relieved forging. (h) Heat treat to 1050 °C (1922 °F) hold for 1 h. (j) Heat treat to 1080 °C (1976 °F) hold for 8 h, plus 700 °C (1290 °F) hold for 16 h. (k) Heat treat to 1150 °C (2100 °F) hold for 4 h, plus 1050 °C (1920 °F) hold for 16 h, plus 850 °C (1560 °F) hold for 16 h. (m) Heat treat to 1190 °C (2175 °F) hold for 1.5 h, plus 1100 °C (2010 °F) hold for 6 h. (n) Heat treat to 1150 °C (2100 °F) hold for 2 h, W.Q., plus 800 °C (1475 °F) hold for 8 h.

tension overloads might occur. Generally, however, materials for high-temperature components are selected on the basis of creep and stress-rupture characteristics.

FATIGUE STRENGTH

Fatigue strength is a very important mechanical property characteristic. It has been estimated that about 90% of all engineering

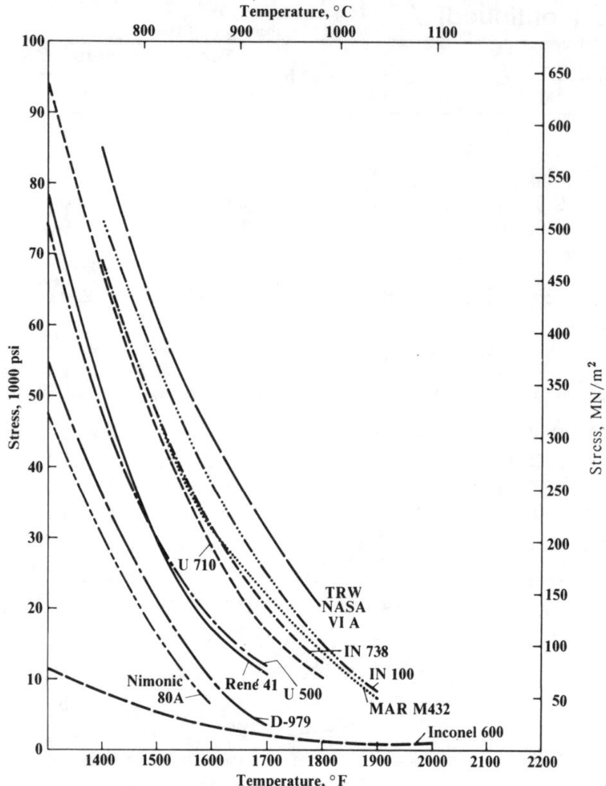

Fig. 4.1 Stress vs temperature curves for rupture in 1000 h for select nickel-base alloys

structures fail because of fatigue. This is due to the fact that cyclic loads in service, particularly those from resonant vibrations, are more difficult to predict than steady loads, and the behavior of engineering materials relative to cyclic loads is very sensitive to their microstructural and macrostructural features.

Fatigue fractures, like steady-stress fractures, are classified as low-temperature, high-temperature, or mixed, depending on the fracture path. Low-temperature fractures usually are characterized by transgranular initiation and propagation, whereas high-temperature fractures are predominantly intergranular. The transition from low- to high-temperature behavior at or near the equicohesive point is a function of temperature, stress, strain rate, and alloy type. These three modes of fatigue fracture are represented in Fig. 4.6.

Low-temperature fatigue is characterized by an induction stage, during which slip bands are produced under stress, which thus develop into microcracks. The microcracks propagate along slip planes toward the interior, and the linking together of these microcracks leads to the formation of a macrocrack. This is followed by a period of slow propagation along a noncrystallographic plane normal to the direction of the principal applied stress, finally ending in rapid crack propagation and fracture.

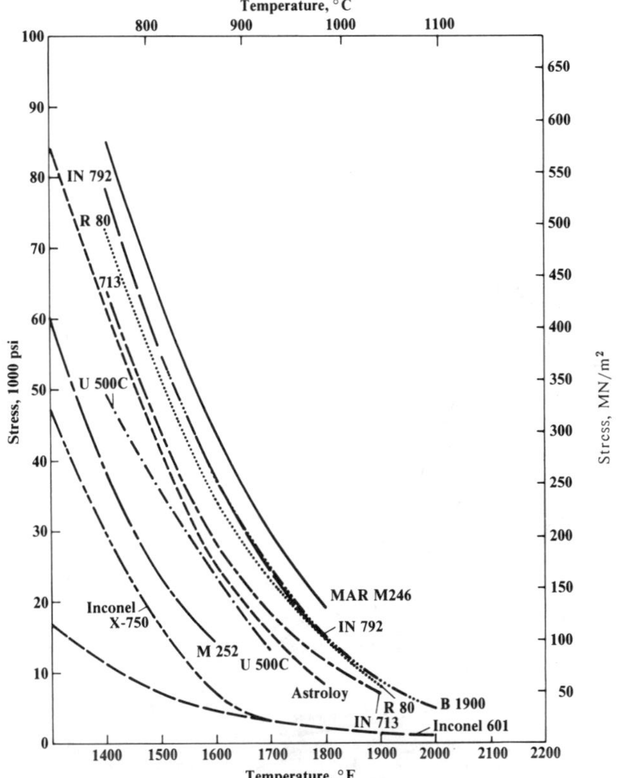

Fig. 4.2 Stress vs temperature curves for rupture in 1000 h for select nickel-base alloys

Solid-solution-strengthened superalloys should be expected to have increased resistance to fatigue cracking through increased resistance to slip and enhanced strain-hardening capacity. Precipitation strengthening would be expected to improve fatigue properties under ideal conditions, because a uniform, stable dispersion of a second phase would increase strain-hardening capacity and would tend to disperse slip. This has been shown to be true with thoria-dispersion-strengthened nickel alloys. In the case of precipitation-strengthened superalloys, however, the precipitated phases are neither uniform, nor are they stable at elevated temperatures. The result is a concentration rather than dispersion of slips and reduced fatigue-to-static-strength ratios. Of course, precipitation strengthening raises the static strength, thereby producing a net increase in fatigue strength despite the reduction in strength ratio. This effect is shown in Table 4.6 for several Nimonic alloys.

Inclusions and voids have a significant influence on the initiation of fatigue cracking and reduction in fatigue strength. The presence of massive intragranular carbides has been shown to reduce fatigue life by up to 50%. The presence of nonmetallic inclusions and geometrical discontinuities often is associated with fatigue failure.

Fatigue at elevated temperatures shows many of the characteristics of creep rupture. As temperature is increased, slip band cracking is

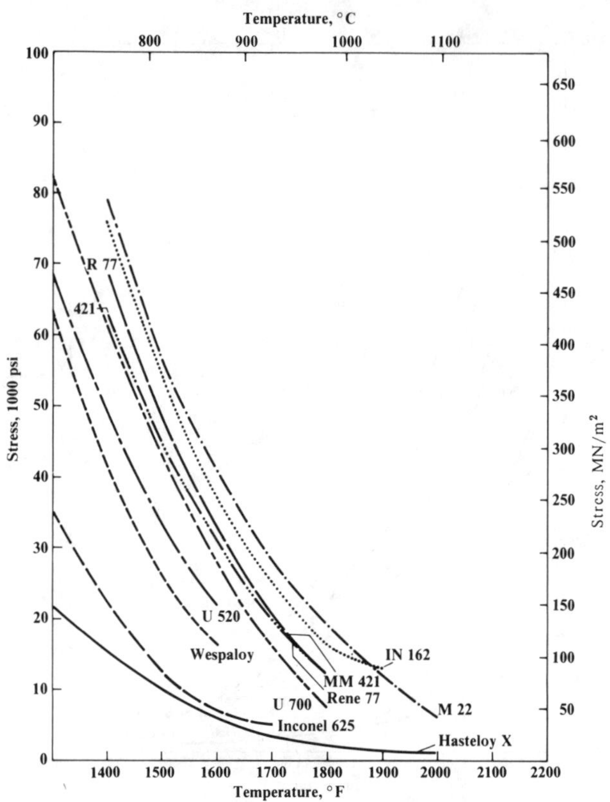

Fig. 4.3 Stress vs temperature curves for rupture in 1000 h for select nickel-base alloys

suppressed in favor of intergranular fracture. In general, high-temperature fatigue may be described as a cyclic creep-rupture process. Thermal fatigue is a particular type of high-temperature fatigue involving heating and cooling cycles during which the component is heated and cooled nonuniformly. Such nonuniform heating produces internal stresses that are compressive in areas that are hotter than the average temperature and produces stresses that are tensile in areas that are cooler than average. During nonuniform cooling, the areas stressed in compression during heating are stressed in tension, and vice versa. When these stresses are of sufficient magnitude and if the cycle is repeated a sufficient number of times, localized fracture results, which is termed a "thermal-fatigue crack."

Although thermal fatigue generally is considered a high-temperature phenomenon, similar to creep-rupture and fatigue failure, it may be described as low temperature or high temperature, depending on whether the fracture path is transgranular or intergranular. Turbine vanes in aircraft jet engines are examples of components that are subjected to thermal fatigue. During acceleration, the leading and trailing edges of turbine vanes heat more quickly and expand more than the cooler mid-chord region. On deceleration, the leading and trailing edges cool more rapidly than the center region. This alternating cycle often results in thermal fatigue cracking of the leading and trailing edges, as shown in Fig. 4.7.

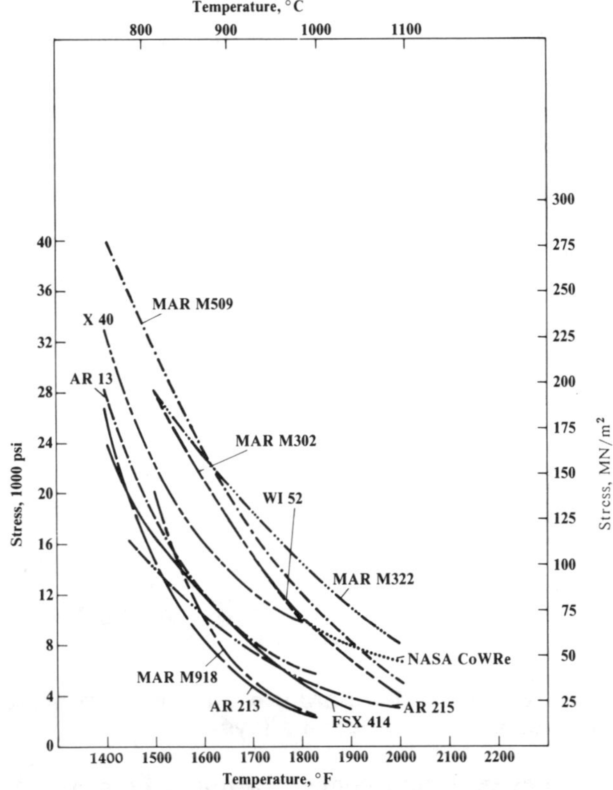

Fig. 4.4 Stress vs temperature curves for rupture in 1000 h for select cobalt-base alloys

CORROSION/OXIDATION RESISTANCE

Superalloys must be able to withstand the deteriorating effects of service atmosphere, corrosion, hot corrosion, and oxidation. From the standpoint of resisting environmental attack, the most important alloying element is chromium, which is present in significant amounts in all superalloys. Because of the presence of chromium, superalloys generally have satisfactory corrosion resistance for many applications.

Resistance to oxidizing environments at high temperatures is an important requisite for superalloys. Early compositions like Nimonic 80A, M-252, A-286, X-40, Rene' 41, V-500, Astroloy, and Waspaloy contained high chromium contents so they possessed sufficient inherent oxidation resistance for the service demanded. However, the service temperatures for superalloys steadily increased, and chromium contents have decreased to obtain the necessary high-temperature strength. Consequently, in many applications, oxidation-resistant coatings are used. (Protective coatings are discussed in Chapter 15.)

The primary concern for superalloy oxidation is its influence on component life by reducing the load-bearing cross section and

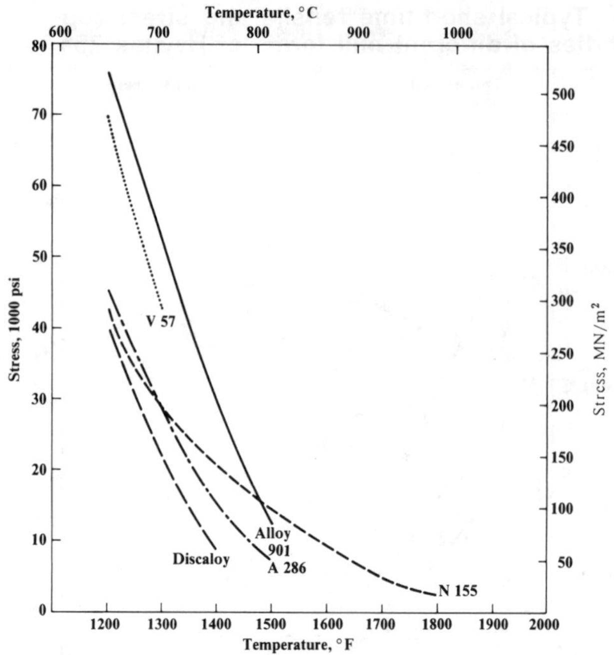

Fig. 4.5 Stress vs temperature curves for rupture in 1000 h for select iron-base alloys

introducing sources of stress concentration, which are detrimental to fatigue strength. These effects occur by surface scaling, internal oxidation, oxide spalling, and oxide vaporization. Superalloy oxidation resistance is enhanced by the formation of protective oxide scales, particularly Cr_2O_3 and Al_2O_3. It follows, therefore, that compositions rich in chromium and aluminum show good oxidation resistance.

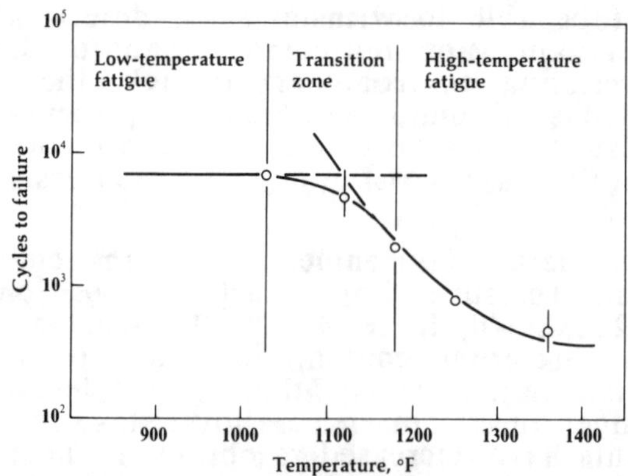

Fig. 4.6 Fatigue data for L-605 tested at a constant total strain range (0.009) and frequency (0.33 Hz) showing the low-temperature and high-temperature fatigue regimes and the transition zone

Table 4.4 Typical short-time tensile and stress-rupture properties of different mill forms of Haynes 25

Product form	Tensile strength MPa	Tensile strength ksi	Yield strength MPa	Yield strength ksi	Elongation, %
At 540 °C (1000 °F)					
Bar	750	109	...	...	71
Sheet	690	100	...	...	68
Tube	710	103	315	46	12
At 650 °C (1200 °F)					
Bar	670	97	...	...	37
Sheet	515	75	...	...	25
Tube	620	90	275	40	12
At 815 °C (1500 °F)					
Bar	455	66	...	...	24
Sheet	345	50	...	...	15
Tube	345	50	170	25	25

	Stress for rupture in: 100 h		450 h		1000 h	
	MPa	ksi	MPa	ksi	MPa	ksi
At 790 °C (1350 °F)						
Bar	255	37	...	...	...	...
Tube	255	37	...	...	...	...
At 815 °C (1500 °F)						
Bar	150	22	...	...	117	17
Sheet	150	22	...	...	121	17.5
At 870 °C (1600 °F)						
Bar	...	...	...	...	86	12.6
Sheet	...	...	82	12	...	...

Note: Bar properties were determined from standard 13-mm (0.5-in.) diameter tension-test bars. Sheet stock was 1.0 to 1.8 mm (0.040 to 0.070 in.) thick. Tube stock was 9.5-mm (0.375-in.) OD by 0.7-mm (0.028-in.) wall thickness, and was tested in full section.

Sulfidation is a specific hot corrosion mechanism that attacks many alloys when sulfur and salt are introduced in the service environment. Alloys high in chromium content (20% or more) have good resistance to sulfidation. Other superalloys may require coatings to combat this attack.

Generalizations about comparative oxidation and hot corrosion resistance of nickel- and cobalt-base superalloys must be treated with some caution, as there are wide ranges of resistance within each alloy group, and seemingly small variations in service conditions can vastly affect results, as shown in Fig. 4.8. Figure 4.9 illustrates hot corrosion resistance of a select group of nickel-base superalloys, and Fig. 4.10 presents oxidation data for select oxide-dispersion-strengthened (ODS) alloys. The corrosion resistance (oxidation and hot corrosion) of a number of alloys is illustrated in Fig. 4.8. Figure 4.11 shows the criterion used to make the evaluation in laboratory and service samples.

Primary factors involved in superalloy selection are cost, performance, and durability of components. It is necessary to achieve satisfactory performance and durability through alloy properties, but

Table 4.5 Effect of cold rolling on elevated temperature properties of Haynes 25 sheet

Cold reduction, %	Temperature		Tensile strength		0.2% yield strength		Rupture time, h, at 130 MPa (19 ksi)
	°C	°F	MPa	ksi	MPa	ksi	
0	540	1 000	690	100	...	...	...
0	870	1 600	310	45	...	...	30
10	540	1 000	725	105	525	76	...
10	870	1 600	425	62	325	47	35
15	540	1 000	930	135	745	108	...
15	870	1 600	485	70	345	50	30
20	540	1 000	1 070	155	930	135	...
20	870	1 600	515	75	415	60	15

Note: Sheet 14 mm (0.054 in.) thick was solution treated at 1220 °C (2225 °F) before cold rolling.

Table 4.6 Fatigue strengths of Nimonic alloys at 750 °C (1380 °F)

Alloy	Al + Ti, wt%	Fatigue strength(a), MPa	Fatigue strength(a), tons/in.^{2}	Fatigue/ultimate strength ratio
Nimonic 75	0.5	165	12	0.67
Nimonic 90	3.8	372	27	0.61
Nimonic 95	4.8	386	28	0.60
Nimonic 105	6.5	414	30	0.52

(a) At 10^6 cycles.

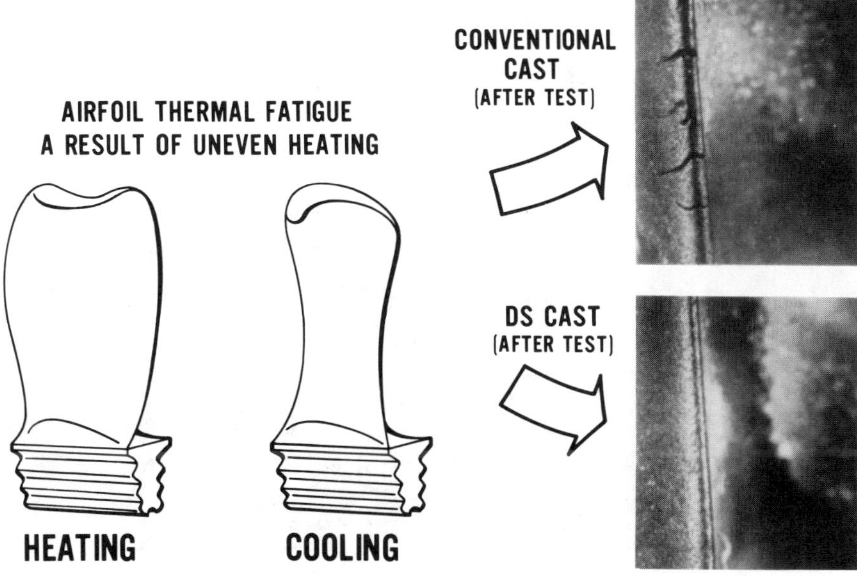

Fig. 4.7 Thermal fatigue cracks in jet engine turbine vane

at the lowest cost possible. Cost includes, in addition to the direct cost, the availability and repairability of the material - two factors that directly affect operating cost. Mechanical and physical properties, as well as operating atmosphere, relate to performance and durability. Materials must be selected on the basis of adequate strength and stability at operating temperature and resistance to fatigue and environmental damage. The materials selection exercise is, in effect, a trade study for each part. Because the aircraft turbine engine is the largest consumer of superalloys, a few examples of superalloy engine parts are presented below.

Hot Disk Applications

The primary design criteria for hot disk applications are yield, burst, fatigue, and creep strengths. Creep plays a significant role because of increasing operating temperature. When the historic progression of alloy properties is compared with compression-exit-temperature increases, there is a suggestion that conventional superalloys are

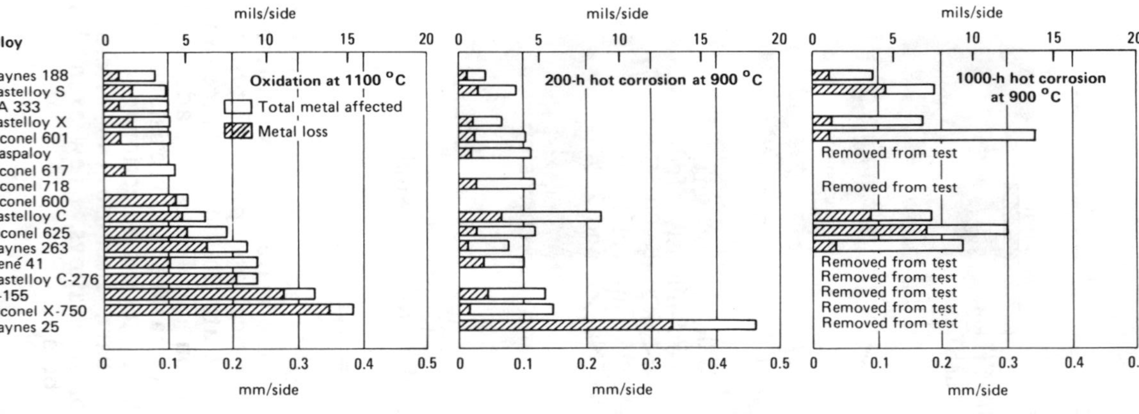

Left: 100-h dynamic oxidation resistance at 1100 °C (2000 °F): tested in an environment created by burning No. 2 fuel oil (0.3 to 0.45% S) with an air-to-fuel ratio of 45:1 to 55:1; thermal shock frequency was 2 per hour. Center: 200-h hot corrosion resistance at 900 °C (1650 °F): tested according to ASTM D-665 in an environment created by burning No. 2 fuel oil (0.3 to 0.45% S), to which 50 ppm salt were added, using an air-to-fuel ratio to 30:1; thermal shock frequency was 1 per hour. Right: 1000-h hot corrosion resistance at 900 °C (1650 °F): conditions same as for center graph, except salt concentration was only 5 ppm. The following alloys were removed from test due to excessive attack and are listed in order of decreasing resistance: N-155 (601 h), Inconel 718 (601 h), Waspaloy (700 h), Rene′ 41 (700 h), Inconel X-750 (700 h), Haynes 25 (142 h), Hastelloy C-276 (296 h); all seven were judged inferior to Inconel 601.

Fig. 4.8 Elevated temperature corrosion of select superalloys

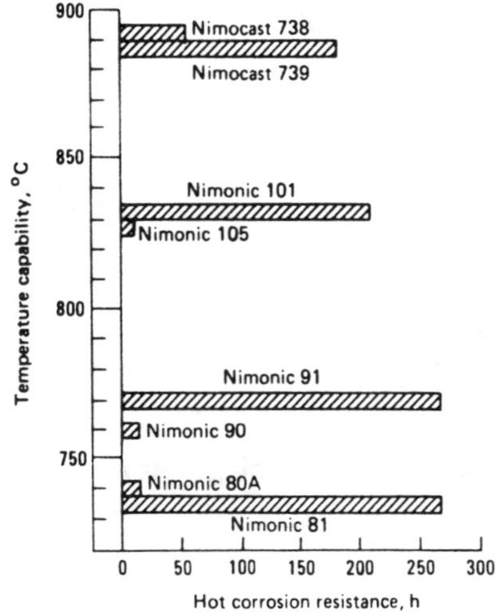

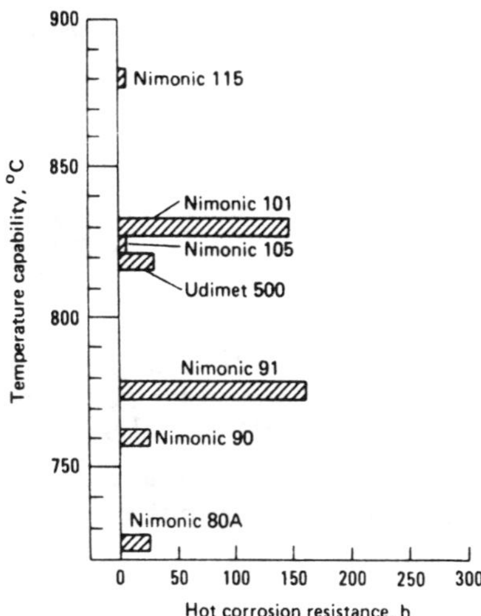

Temperature capabilities based on 10,000-h rupture life at 120 MPa (17.4 ksi). Hot corrosion resistance based on time to a weight loss of: left, 1 kg/m^2 (100 mg/cm^2) in continuously fed mixture of 75% Na_2SO_4 and 25% NaCl at 900 °C (1650 °F); right, 180 g/m^2 (18 mg/cm^2) in continuously fed mixture of 75% Na_2SO_4 and 25% NaCl at 850 °C (1560 °F).

Fig. 4.9 Temperature capabilities and hot corrosion of several Nimonic alloys

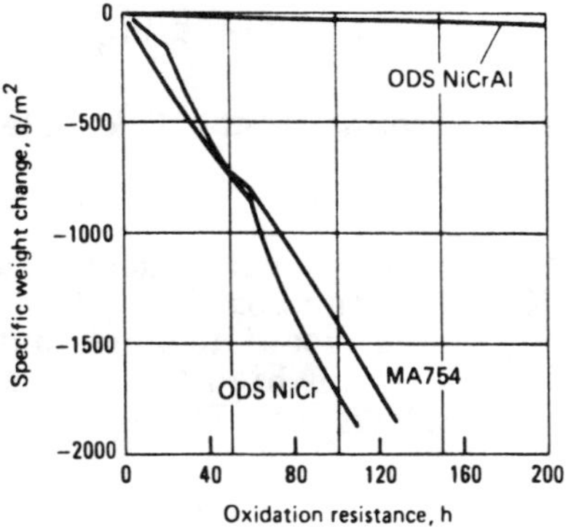

Tested in Mach 1 burner rig at 1200 °C (2200 °F); samples cooled to room temperature once each hour.

Fig. 4.10 Oxidation resistance of select oxide-dispersion-strengthened alloys

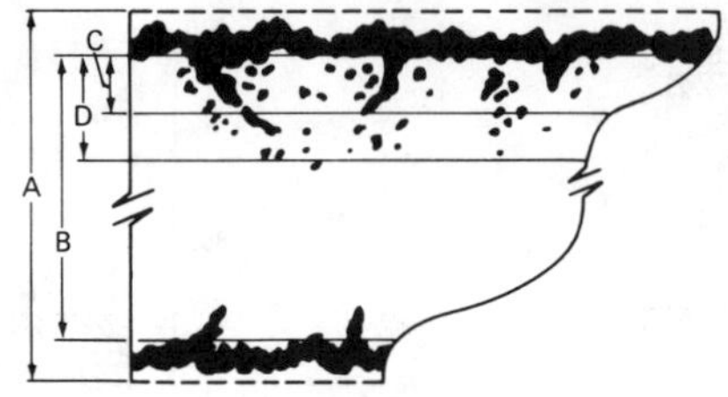

Metal loss per side = (A-B)/2
Continuous penetration per side = C
Maximum penetration per side = D
Total metal affected per side = [(A-B)/2] + D

Fig. 4.11 Schematic of metallographic technique for measuring hot corrosion damage

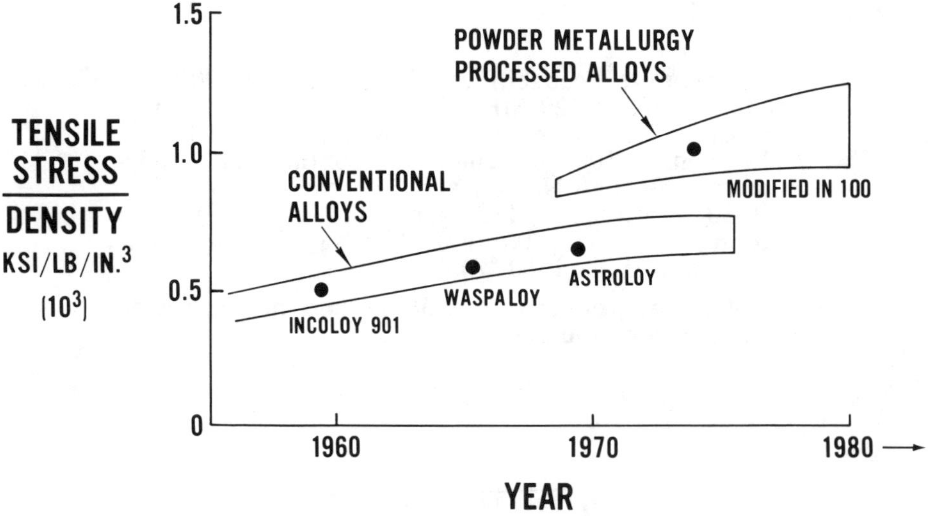

Fig. 4.12 Trends in hot disk materials

beginning to top-out. Use of powder metallurgy (P/M) processing with modified IN-100 alloy compositions, however, provides combined improvement. It is significant to note that use of IN-100 type compositions is based not only on high-temperature strength, but also on favorable density characteristics. Trends in hot disk materials are shown in Fig. 4.12.

Burner Applications

The burner is a simple device in which tremendous amounts of heat are released in a very small space. This results in high local temperatures. Primary design criteria for the burner are thermal fatigue resistance, tolerance for local overtemperature conditions,

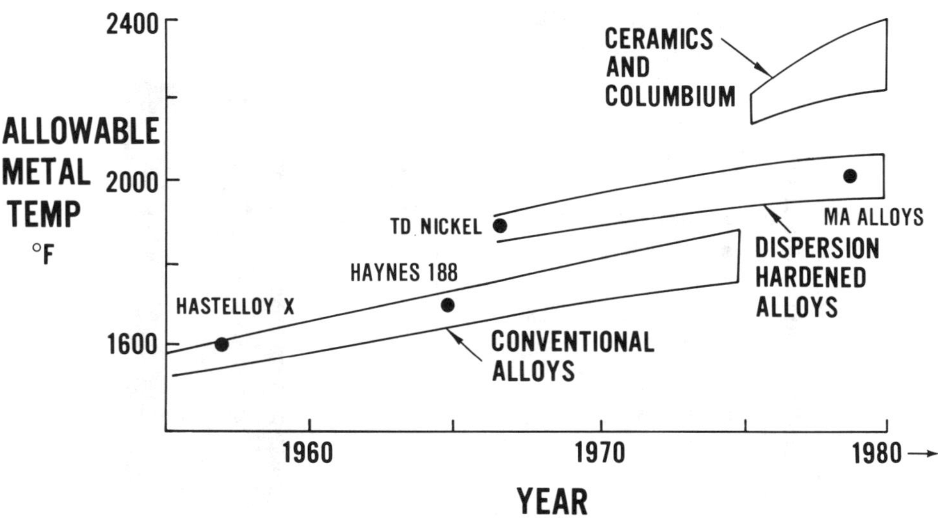

Fig. 4.13 Trends in burner materials

resistance to corrosion/erosion, and good creep strength. There has been some increase in alloy temperature capability over the years, but not enough to keep pace with the increase in combustor temperatures. The gap is filled, of course, by design innovations, notably air cooling. Trends in burner materials are shown in Fig. 4.13.

Hot Airfoil Applications

For high-temperature airfoil applications, the important material properties are creep strength, stress-rupture strength, and resistance to thermal fatigue and environmental damage of blades (the rotating airfoils). Repair weldability is also a factor for vanes (static airfoils). Figure 4.14 lists the design criteria for the rotating turbine blade airfoil. Results of excessive creep of turbine blades are characterized by necking, shroud rubbing, and bowing. Turbine airfoils can fail in other ways before excessive creep has been experienced. Thermal fatigue cracking of the airfoil can result in failure caused either by stress-rupture or mechanical fatigue. Turbine airfoils may also suffer deterioration through corrosion (sulfidation) and oxidation attack.

Historically, the nickel-base precipitation-strengthened superalloys have performed well in turbine airfoil designs. Materials improvements in temperature capability, however, have not kept pace with engine temperature increases. Thus, airfoils have been forced to move from forged hardware to cast hardware to air cooled designs to provide satisfactory service durability. Even with cooling designs, strength requirements have increased, and improved alloy capability has been required. This improved alloy capability has come about through the use of directionally solidified castings, in which aligned axial grains are produced instead of random grains, or single-crystal

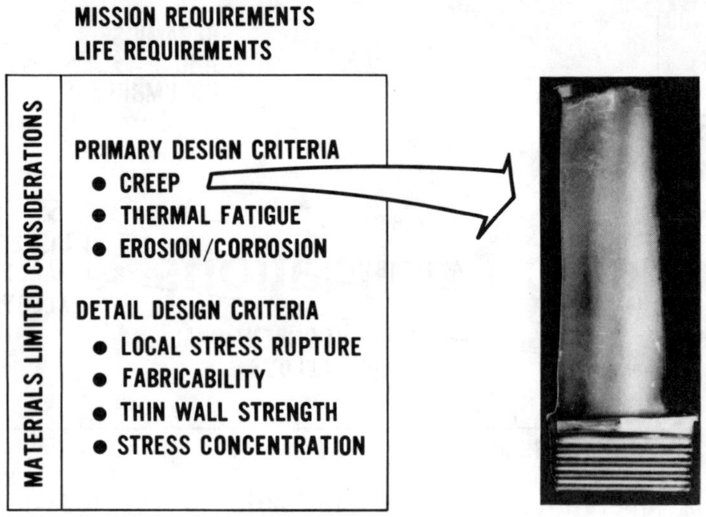

Fig. 4.14 Considerations for high-temperature airfoil design

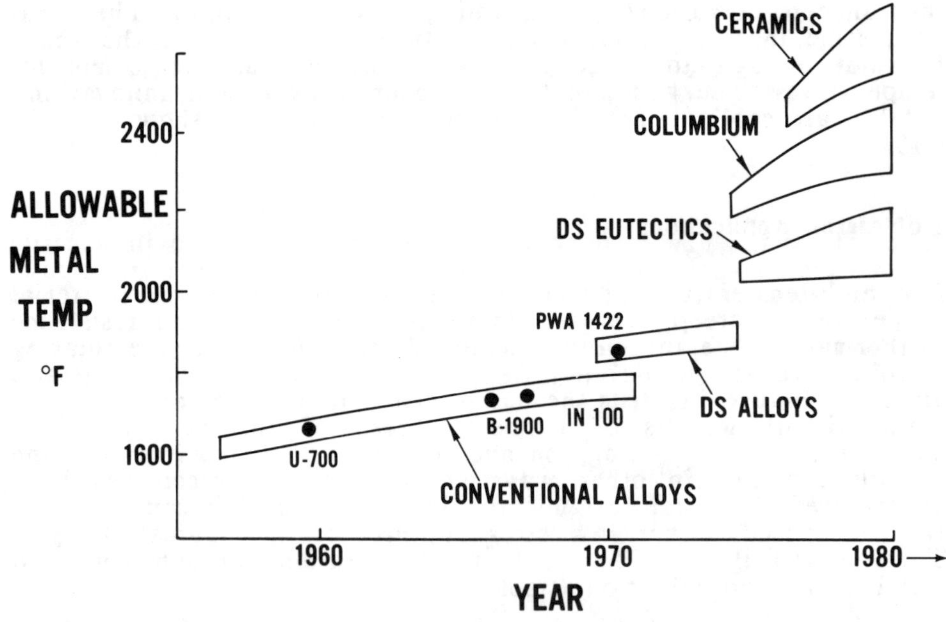

Fig. 4.15 Trends in turbine airfoil materials

castings, in which all grain boundaries are eliminated. The directionally solidified and single-crystal processes improve the thermal fatigue and ductility projection of alloys, thereby allowing use of high-strength compositions that are excessively brittle when used in the conventional cast structure. Trends in turbine airfoil materials are shown in Fig. 4.15.

Chapter 5

Cryogenic Applications

INTRODUCTION

Although superalloys generally are designed for high-temperature applications, certain nickel-base superalloy-type compositions possess excellent mechanical characteristics at cryogenic temperatures. All structural metals undergo changes in properties when cooled from room temperature to subzero temperatures below 0 °C (32 °F). The greatest changes in properties occur when the metal is cooled to very low temperatures near the boiling points of liquid hydrogen and liquid helium. However, even at the less severe subzero temperatures encountered in arctic regions, where the lowest temperature recorded has been –71 °C (–96 °F), many materials become embrittled.

MATERIALS SELECTION

The effects of subzero temperatures must be considered in selection of materials for aircraft, missiles, and space vehicles that are exposed to the temperatures of upper altitudes and outer space. These structures are "weight limited," so they must be fabricated from materials with high strength-to-weight ratios both at room temperature and at subzero temperatures. At the same time, these materials are required to retain high levels of fracture toughness at all exposure temperatures for "fail safe" service.

Property requirements are critical in selecting materials for welded structures such as liquid oxygen, liquid hydrogen, and liquid helium tankage and associated piping and fittings for rockets and launch vehicles. Among these requirements are minimum weight, high toughness of the base metal at cryogenic temperatures, and high strength and toughness of the welded joints. Materials that have been used successfully for these structures include several high-nickel alloys.

Certain metals, alloys, and compounds become superconductors at temperatures below about –260 °C (–436 °F). To achieve these temperatures, all superconducting devices must be cooled with liquid helium. Therefore, structural materials selected for cryogenic components of superconducting machinery, magnets, and transmission systems must be suitable for use at liquid helium temperature. Furthermore, these components are subjected to high stresses in

Table 5.1 Nominal compositions of high-nickel alloys

UNS number	Alloy designation	Nominal composition, % Ni	Cr	Fe	Mn	Si	C	Others
N05500	Monel K-500	Rem	...	1.0	0.6	0.15	0.15	29.5Cu, 2.8Al, 0.5Ti
	Hastelloy B	Rem	0.6	5.0	0.8	0.7	0.1	2.5Co, 28.0Mo, 0.2-0.6V
	Hastelloy C	Rem	15.5	5.5	0.5	0.5	0.07	1.5Co, 16 Mo, 4.0W
N06600	Inconel 600	Rem	15.8	7.2	0.2	0.2	0.04	0.10 Cu
N09706	Inconel 706	39 to 44	16	Rem	0.10	0.10	0.04	0.35 max Al, 3.0 (Nb + Ta), 1.7Ti
N07718	Inconel 718	Rem	18.6	18.5	...	...	0.04	0.4Al, 0.9Ti, 5.0Nb, 3.1Mo
N07750	Inconel X-750	Rem	15.0	6.8	0.7	...	0.04	0.8Al, 2.5Ti, 0.85Nb
	Invar 36	36	...	Rem	...	...	...	...

service, and thus, safeguards must be taken to minimize service failures. To obtain the required strength and toughness along with a reasonable degree of fabricability, certain nickel-base superalloys usually are designated for highly stressed components of structures that will be cooled with liquid helium.

Nickel is a face-centered cubic (fcc) metal that retains good ductility and toughness at subzero temperatures. Unalloyed nickel is low in strength and has only limited usage at subzero temperatures. However, several nickel-base alloys, including some superalloys, have been evaluated at cryogenic temperatures. Compositions of select high-nickel alloys are given in Table 5.1. Some of these alloys exhibit excellent combinations of strength, ductility, and toughness over the entire range of subzero temperatures. These alloys have been selected for some of the most critical structural components in recent designs for large superconducting motors and generators.

They also are suitable for service at elevated temperatures and may be used in applications involving exposure to both subzero and elevated temperatures.

MECHANICAL PROPERTIES

Mechanical property data on tensile strength, fracture toughness, fatigue-crack growth rate, fatigue strength, Young's modulus, and Poisson's ratio at room temperature and at subzero temperatures are presented in Tables 5.2 and 5.3 for several alloys in various mill forms and for some weldments. The values presented have been averaged from available property data and represent state-of-the-art information. Because these test data are limited, the values represent only composite averages and are not statistical means or design values.

Tensile Properties

Typical tensile properties of high-nickel alloys, and of high-nickel alloy weldments, at room temperature and at subzero temperatures are presented in Tables 5.2 and 5.3. Typical yield strengths at room temperature for the heat treated alloys range from 703 to 1172 MPa (102 to 170 ksi). Maximum yield strength at -269 °C (-452 °F) is 1406 MPa (204 ksi). All of the heat treated alloys in Table 5.2, and some of the cold rolled alloys, retain good ductility at the lowest testing temperature. Notch tensile data were not available for some of these alloys, but the available data indicate that they also retain good notch toughness at the lowest testing temperatures. Ductility of heat treated weldments is lower, but is not affected by extended exposure to subzero temperatures.

Values of Young's modulus at subzero temperatures for Inconel 718, Inconel X-750, and Inconel 600 are given in Fig. 5.1 and 5.2. Values of Poisson's ratio at subzero temperatures for Inconel 600 and Inconel X-750 are presented in Fig. 5.3.

Table 5.2 Typical tensile properties of nickels and high-nickel alloys

Temperature		Tensile strength		Yield strength		Elongation,	Reduction in area	Notch tensile strength(a)		Young's modulus	
°C	°F	MPa	ksi	MPa	ksi	%	%	MPa	ksi	GPa	10^6 psi
Hastelloy B sheet, cold rolled 40%, longitudinal orientation											
24	75	1320	191	1220	177	3	...	...	...	...	...
−73	−100	1530	222	1430	207	5	...	...	...	...	...
−196	−320	1570	228	1430	208	12	...	...	...	...	...
−253	−423	1950	283	1650	240	16	...	...	...	...	...
Hastelloy C sheet, cold rolled 20%, longitudinal orientation											
24	75	1140	165	1000	145	13	...	1250	182	...	...
−196	−320	1520	220	1280	186	32	...	1560	226	...	...
−253	−423	1740	252	1380	200	33	...	1690	245	...	...
Inconel 600 sheet, hard, cold rolled, longitudinal orientation											
24	75	910	132	885	128	4	...	...	...	...	...
−253	−423	1210	176	910	132	22	...	...	...	...	...
Inconel 600 bar, cold drawn, longitudinal orientation											
24	75	940	136	890	129	15	56	1230	179	170	25
−78	−108	985	143	910	132	20	58	...	...	...	...
−196	−320	1160	168	1030	150	26	62	...	...	...	...
−253	−423	1250	181	1100	160	30	56	...	...	...	...
−257	−430	1280	186	1210	176	20	56	1530	222	220	32
Inconel 706 forged billets(b)											
24	75	1260	183	1050	152	24	33	1880	272	...	...
−196	−320	1570	228	1200	174	29	33	2170	315	...	...
−269	−452	1680	243	1250	181	30	33	2250	326	...	...

(a) K_t = 10 for Inconel 706 forged billets, Inconel 718 sheet, and Inconel X-750 forged billets; K_t = 6.3 for Hastelloy C sheet, Inconel 718 forgings, and Inconel X-750; K_t = 6.4 for Inconel 600 sheet. (b) Aged 1 h at 980 °C (1800 °F), air cooled, 8 h at 730 °C (1350 °F), furnace cooled to 620 °C (1150 °F), hold 8 h, air cooled. (c) Aged 1 h at 955 °C (1750 °F), air cooled, 8 h at 720 °C (1325 °F), furnace cooled to 620 °C (1150 °F), hold 10 h, air cooled. (d) Aged 3/4 h at 980 °C (1800 °F), air cooled, 8 h at 720 °C (1325 °F), furnace cooled to 620 °C (1150 °F), hold 10 h, air cooled. (e) Annealed and aged 20 h at 700 °C (1300 °F), air cooled.

(continued)

Table 5.2 (continued)

Temperature		Tensile strength		Yield strength		Elongation,	Reduction in area	Notch tensile strength(a)		Young's modulus	
°C	°F	MPa	ksi	MPa	ksi	%	%	MPa	ksi	GPa	10^6 psi
Inconel 718 sheet, longitudinal orientation(c)											
24	75	1330	193	1090	158	18	...	1330	193	205	29.9
−78	−108	1490	216	1190	172	17	...	1470	213	220	31.7
−196	−320	1730	251	1310	190	21	...	1560	226	225	32.5
−253	−423	1740	252	1340	194	16	...	1500	217	225	32.6
Inconel 718 sheet, transverse orientation(c)											
24	75	1320	192	1100	159	18	...	1300	188	200	28.8
−78	−108	1480	214	1210	176	12	...	1450	210	210	30.8
−196	−320	1700	246	1300	189	21	...	1500	217	230	33.5
−253	−423	1770	256	1370	198	16	...	1500	218	240	34.5
Inconel 718 bar, longitudinal orientation(d)											
24	75	1410	204	1170	170	15	18	...	...	...	...
−196	−320	1650	239	1340	197	21	20	...	...	...	...
−269	−452	1810	263	1410	204	21	20	...	...	...	...
Inconel 718 forgings, longitudinal orientation(d)											
24	75	1340	194	1150	167	24	35	2030	295	...	...
−78	−108	1350	196	1190	172	29	45	2170	314	...	...
−196	−320	1630	237	1300	188	26	34	2350	341	...	...
−253	−423	1680	244	1320	192	28	42	2390	347	...	...
−269	−452	1810	263	1410	204	21	20	...	...	...	...
Inconel 718 forgings, transverse orientation(d)											
24	75	1290	187	1150	167	18	28	1930	280	...	...
−253	−423	1740	253	1350	196	24	30	2300	333	...	...
Inconel 718 forgings, S-T orientation(d)											
24	75	1290	187	1140	166	17	23	1860	270	...	...
−253	−423	1630	237	1340	195	14	12	1970	286	...	...

(a) K_t = 10 for Inconel 706 forged billets, Inconel 718 sheet, and Inconel X-750 forged billets; K_t = 6.3 for Hastelloy C sheet, Inconel 718 forgings, and Inconel X-750; K_t = 6.4 for Inconel 600 sheet. (b) Aged 1 h at 980 °C (1800 °F), air cooled, 8 h at 730 °C (1350 °F), furnace cooled to 620 °C (1150 °F), hold 8 h, air cooled. (c) Aged 1 h at 955 °C (1750 °F), air cooled, 8 h at 720 °C (1325 °F), furnace cooled to 620 °C (1150 °F), hold 10 h, air cooled. (d) Aged 3/4 h at 980 °C (1800 °F), air cooled, 8 h at 720 °C (1325 °F), furnace cooled to 620 °C (1150 °F), hold 10 h, air cooled. (e) Annealed and aged 20 h at 700 °C (1300 °F), air cooled.

(continued)

Table 5.2 (continued)

Temperature		Tensile strength		Yield strength		Elongation,	Reduction in area	Notch tensile strength(a)		Young's modulus	
°C	°F	MPa	ksi	MPa	ksi	%	%	MPa	ksi	GPa	10^6 psi
Inconel X-750 sheet, longitudinal orientation(e)											
24	75	1220	177	815	118	24	...	1120	162	210	30.4
−78	−108	1320	192	875	127	28	...	1200	174	...	...
−196	−320	1500	217	905	131	32	...	1270	184	225	32.4
−253	−423	1590	230	940	136	32	...	1370	199	...	...
Inconel X-750 sheet, transverse orientation(e)											
24	75	1230	178	850	123	25	...	1160	168	...	...
−78	−108	1340	194	925	134	26	...	1210	175	...	...
−196	−320	1500	217	950	138	32	...	1270	184	...	...
−253	−423	1630	236	985	143	32	...	1390	201	...	...
Inconel X-750 bar, longitudinal orientation(e)											
24	75	1340	194	985	143	25	49	...	...	...	...
−196	−320	1570	228	1050	152	32	45	...	...	...	...
−253	−423	1700	246	1090	158	33	42	...	...	...	...
−257	−430	1720	249	1080	157	33	46	...	...	...	...
Inconel X-750 forged billet(b)											
24	75	985	143	665	96.2	18	18	1200	174	...	...
−196	−320	1090	158	770	112	16	14	1340	195	...	...
−269	−452	1020	148	735	107	14	13	1410	205	...	...
Invar 36 bar, cold drawn 12 to 15%, longitudinal orientation											
24	75	650	94	625	91	21	62	...	...	...	...
−78	−108	785	114	725	105	29	60	...	...	...	...
−196	−320	1080	156	915	133	27	61	...	...	...	...
−253	−423	1190	172	1120	162	23	58	...	...	...	...
−269	−452	1230	178	1110	161	20	52	...	...	...	...

(a) K_t = 10 for Inconel 706 forged billets, Inconel 718 sheet, and Inconel X-750 forged billets; K_t = 6.3 for Hastelloy C sheet, Inconel 718 forgings, and Inconel X-750; K_t = 6.4 for Inconel 600 sheet. (b) Aged 1 h at 980 °C (1800 °F), air cooled, 8 h at 730 °C (1350 °F), furnace cooled to 620 °C (1150 °F), hold 8 h, air cooled. (c) Aged 1 h at 955 °C (1750 °F), air cooled, 8 h at 720 °C (1325 °F), furnace cooled to 620 °C (1150 °F), hold 10 h, air cooled. (d) Aged 3/4 h at 980 °C (1800 °F), air cooled, 8 h at 720 °C (1325 °F), furnace cooled to 620 °C (1150 °F), hold 10 h, air cooled. (e) Annealed and aged 20 h at 700 °C (1300 °F), air cooled.

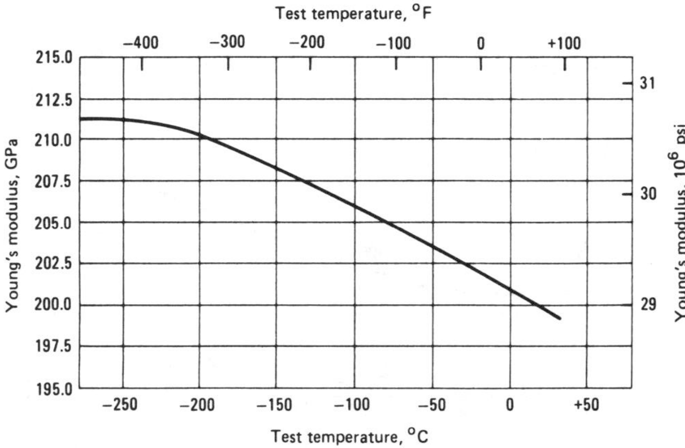

Fig. 5.1 Young's modulus for Inconel 718 as determined ultrasonically

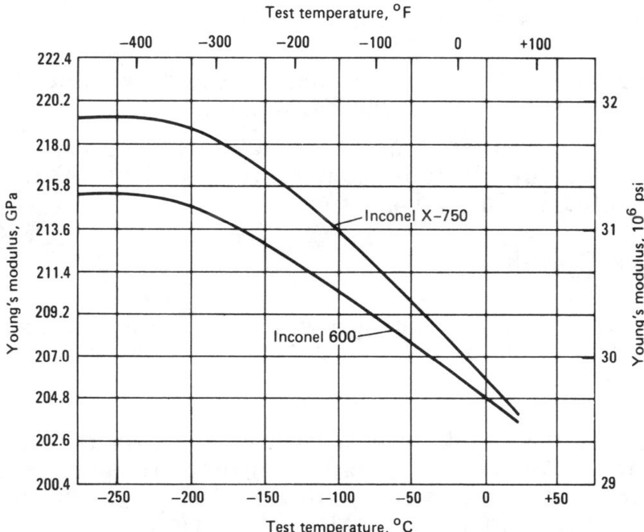

Fig. 5.2 Young's modulus for Inconel 600 and Inconel X-750 as determined ultrasonically

Fracture Toughness

Fracture toughness of one heat of Inconel X-750 that had been vacuum-induction melted and vacuum-arc remelted (VIM-VAR) was abnormally low because of carbide precipitation at the grain boundaries. This precipitation probably occurred because of insufficient breakdown of the billet during forging. The condition was not evident from results of tensile tests, but it was evident from metallographic examination. The fracture toughness data for the other heats of Inconel X-750 are more representative, and they normally retain a high degree of toughness at temperatures as low as -269 °C (-452 °F). Fracture toughness of the fusion zones and

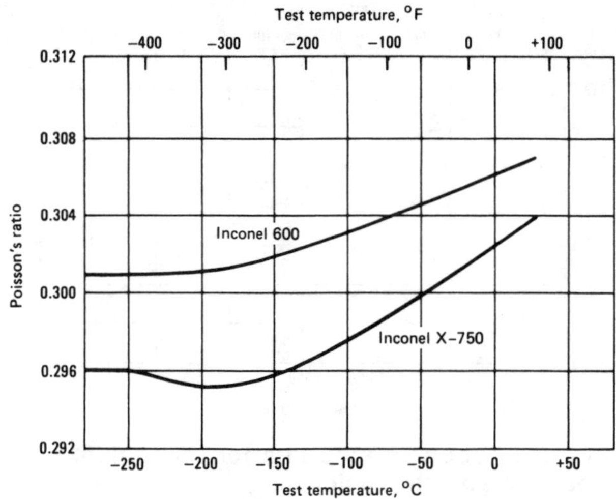

Fig. 5.3 Poisson's ratios for Inconel 600 and Inconel X-750 as determined ultrasonically

heat-affected zones of welds in heat treated weldments, however, tend to be lower than that of the base metal.

Available information on fracture toughness of several high-nickel alloys and weldments at room temperature and at subzero temperatures is presented in Table 5.4.

Fatigue-Crack Growth

Data on rates of fatigue-crack growth for several high-nickel alloys are presented in Table 5.5. These data are for the constants C and n in the equation:

$$da/dN = C(\Delta K)^n$$

where da/dN is the crack-growth rate and ΔK is the stress-intensity-factor range for constant-load-amplitude fatigue tests on precracked specimens. It is not applicable for negative stress ratios. However, the constants in Table 5.5 permit calculation of the fatigue-crack growth rate at any ΔK value within the limits of the range of ΔK.

All of the data in Table 5.5 were obtained at the same stress ratio (R) and at nearly the same frequency. Variations in these conditions may affect fatigue-crack growth rates. However, for these alloys, fatigue-crack growth rates at subzero temperatures are either equal to or lower than the rates at room temperature for the same ΔK values. For design purposes, the use of room-temperature fatigue-crack growth rate data for subzero applications is feasible.

Results of fatigue-life tests at 10^6 cycles on axial and flexural specimens of several high-nickel alloys at room temperature and at subzero temperatures are presented in Table 5.6. These data indicate that, at 10^6 cycles, the fatigue strengths of these alloys are higher at subzero temperatures than at room temperature. Furthermore, specimens with smoother surface finishes had higher fatigue strengths at each testing temperature.

Table 5.3 Typical tensile properties of high-nickel alloy GTAW weldments(a)

Temperature		Tensile strength		Yield strength		Elongation	Reduction in area	Notch tensile strength(b)	
°C	°F	MPa	ksi	MPa	ksi	%	%	MPa	ksi
Inconel 706 forging with 718 filler metal(c)									
24	75	1110	161	1000	145	2	5	1630	236
−196	−320	1300	188	1170	170	4	4	1760	255
−269	−423	1370	199	1220	177	4	6	1880	273
Inconel 718 sheet, no filler metal(d)									
24	75	1320	191	1150	167	5	...	1200	174
−196	−320	1560	226	1290	187	4	...	1320	192
−253	−423	1730	251	1410	205	5	...	1440	209
Inconel 718 forging with 718 filler metal(c)									
24	75	1260	183	2000	159	2	6	1390	202
−196	−320	1430	208	1280	186	2	4	1470	213
−269	−452	1650	239	1280	185	28	33	2280	331
Inconel X-750 sheet with X-750 filler metal(e)									
24	75	1290	187	860	125	22	...	...	...
−78	−108	1340	195	915	133	24	...	...	...
−196	−320	1540	224	945	137	30	...	...	...
−253	−423	1660	241	1020	148	28	...	...	...
Inconel X-750 forged billet with F69 filler metal(d)									
24	75	1100	159	855	124	9	12	1570	228
−196	−320	1110	161	930	135	6	9	1710	248
−269	−452	1120	163	960	139	6	9	1630	236

(a) Base metal has longitudinal orientation. (b) For Inconel 706, Inconel 718 forging with 718 filler, and Inconel X-750 forged billet with F69 filler, $K_t = 10$; for Inconel 718 sheet, no filler, $K_t = 6.3$. (c) Weldment heat treatment: 1 h at 980 °C (1800 °F), air cooled, 8 h 730 °C (1325 °F), furnace cooled to 620 °C (1150 °F), hold 8 h, air cooled. (d) Weldment aged 8 h at 720 °C (1325 °F), furnace cooled to 620 °C (1150 °F), hold 10 h, air cooled. (e) Weldment aged 20 h at 700 °C (1300 °F), air cooled.

Table 5.4 Fracture toughness of high-nickel alloys and weldments

Alloy and condition(a)	Form	Room temperature yield strength		Specimen design	Orientation	Fracture toughness, K_{Ic} or K_{Ic} (J), at: 24 °C (75 °F)		−196 °C (−320 °F)		−269 °C (−452 °F)	
		MPa	ksi			$MPa\sqrt{m}$	$ksi\sqrt{in.}$	$MPa\sqrt{m}$	$ksi\sqrt{in.}$	$MPa\sqrt{m}$	$ksi\sqrt{in.}$
Inconel 706 (VIM-VAR) STDA	Forging	1063	154	CT	C-R.....	134(b)	121(b)	...	...	158(b)	143(b)
Inconel 706 (VIM-VAR) GTA weld, ST/W/STDA	Forging-weldment	1063	154	CT	....	...	...	...	...	58.7(b)(c)	53.0(b)(c)
Inconel 718 STDA	Bar	1172	170	CT	T-S.....	96.3	87.8	103	94.0	112	102
Inconel 718 (VIM-VAR) STDA	Forging	1164	169	CT	C-R.....	61.5(b)	55.6(b)	...	...	75.5(b)	68.2(b)

ST/W/ STDA											
Inconel X-750 (VIM-VAR) STDA	Forging	824	120	CT	C-R	...	...	...	...	76.1(b)	69.2(b)(c)
Inconel X-750 (VIM) STDA	Forging	917	133	CT	C-R	...	...	...	...	145(b)	132(b)
Inconel X-750 (AAM-VAR) STDA	Forging	848	123	CT	C-R	...	...	...	...	237(b)	216(b)
Inconel X-750 (VIM-VAR) ST/W/ STDA	Forging-weldment	824	120	CT	...	...	...	...	...	134(b)(c)(f) 176(b)(c)(g)	122(b)(c)(f) 160(b)(c)(g)

(a) VIM, vacuum induction melted; VAR, vacuum arc remelted; AAM, air arc melted; ST, solution treated; W, welded. STDA for Inconel 706 and Inconel X-750: 980 °C (1800 °F) 1 h, AC, 730 °C (1350 °F) 8 h, FC to 620 °C (1150 °F), hold 8 h, AC. STDA for Inconel 718: 980 °C (1800 °F) 1 h, AC, 720 °C (1325 °F) 8 h, FC to 620 °C (1150 °F), hold 8 h, AC. Filler metals: F-718 for Inconel 706 and Inconel 718; Inco F69 for Inconel X-750. (b) K_{Ic}. (c) Fusion zone. (d) Heat-affected zone. (e) This heat of Inconel X-750 had carbide precipitates at the grain boundaries, which caused abnormally low fracture toughness. (f) Gas tungsten arc weld. (g) Vacuum electron beam weld.

Table 5.5 Fatigue-crack growth rate data for compact specimens of high-nickel alloys

Alloy and condition(a)	Orientation	Frequency, Hz	Stress ratio, R	Test temperature °C	Test temperature °F	C(b) da/dN: mm/cycle ΔK: MPa$\sqrt{m}$	C(b) da/dN: in./cycle ΔK: ksi$\sqrt{in.}$	n	Estimated range for ΔK MPa$\sqrt{m}$	Estimated range for ΔK ksi$\sqrt{in.}$
Inconel 706 forgings	C-R.	10	0.1	24	75	7.42×10^{-11}	4.2×10^{-12}	4.11	22 to 44	20 to 40
VIM-EFR STDA				−196	−320	5.89×10^{-10}	3.18×10^{-11}	3.35	22 to 88	20 to 80
				−269	−452	8.93×10^{-10}	4.72×10^{-11}	3.12	24 to 77	22 to 70
VIM-VAR STDA	C-R.	10	0.1	24	75	1.02×10^{-9}	5.47×10^{-11}	3.27	22 to 53	20 to 48
				−196	−320	2.87×10^{-10}	1.57×10^{-11}	3.5	25 to 77	23 to 70
				−269	−452	2.33×10^{-9}	1.19×10^{-10}	2.77	25 to 88	23 to 80
VIM-VAR, GTA weld	C-R.	10	0.1	24	75	1.02×10^{-10}	6.02×10^{-12}	4.34	22 to 35	20 to 32
+ STDA (FZ)				−269	−452	2.65×10^{-11}	1.57×10^{-12}	4.35	23 to 40	21 to 36
Inconel 718 forging	C-R.	10	0.1	24	75	1.52×10^{-10}	8.79×10^{-12}	4.10	22 to 48	20 to 44
VIM-VAR STDA				−196, −269	−320, −452	7.31×10^{-12}	4.42×10^{-13}	4.55	22 to 55	20 to 50
Inconel 718 forging	T-S.	20-28	0.1	24	75	8×10^{-11}	4.59×10^{-12}	4.0	20 to 70	18 to 64
STDA				−78, −196, −269	−108, −320, −452	4.8×10^{-11}	2.75×10^{-12}	4.0	25 to 90	23 to 82
Inconel X-750 forging	T-S.	20-28	0.1	24	75	2.4×10^{-9}	1.25×10^{-10}	3.0	26 to 80	24 to 73
STDA				−196, −269	−320, −452	6.6×10^{-11}	3.72×10^{-12}	3.8	30 to 92	27 to 84
VIM-VAR STDA	C-R.	10	0.1	24	75	6.34×10^{-15}	4.83×10^{-16}	7.0	23 to 44	21 to 40
				−269	−452	2.44×10^{-17}	2.04×10^{-18}	8.0	33 to 55	30 to 50
AAM-VAR STDA	C-R.	10	0.1	24, −196	75, −320	4.62×10^{-10}	2.52×10^{-11}	3.45	33 to 66	30 to 60
				−269	−452	1.52×10^{-10}	8.26×10^{-12}	3.45	38 to 82	35 to 75
VIM STDA	C-R.	10	0.1	24, −196, −269	75, −320, −452	2.63×10^{-11}	1.57×10^{-12}	4.4	27 to 55	25 to 50

(a) STDA for Inconel 706: 980 °C (1800 °F) 1 h, AC, 730 °C (1350 °F) 8 h, FC to 620 °C (1150 °F), hold at 620 °C (1150 °F) 8 h, AC. STDA for Inconel 718 and Inconel X-750: same as for Inconel 706. (b) Conversion factors for C in equation $da/dN = C\Delta K^n$. If da/dN is in in./cycle and ΔK is in ksi$\sqrt{in.}$, multiply C by $25.4/1.0989^n$ to obtain mm/cycle with ΔK in MPa$\sqrt{m}$. If da/dN is in mm/cycle and ΔK is in MPa $\sqrt{m}$, multiply C by $1.0989^n/25.4$ to obtain in./cycle with ΔK in ksi$\sqrt{in.}$

Table 5.6 Results of fatigue-life tests on unnotched specimens of high-nickel alloys

Alloy and condition(a)	Stressing mode	Stress ratio, R	Cyclic frequency, Hz	Fatigue strengths at 10^6 cycles: 24 °C (75 °F) MPa	24 °C (75 °F) ksi	−196 °C (−320 °F) MPa	−196 °C (−320 °F) ksi	−253 °C (−423 °F) MPa	−253 °C (−423 °F) ksi
K Monel Sheet	Flex(b)	−1.0	...	380	55	395	57	475	69
	Flex(c)	−1.0	...	380	55	470	68	580	84
K Monel bar, STA	Axial	0	28	615	89	800	116	840	122
Inconel X-750 bar, STA	Axial	0	28	745	108	1010	147	1060	154
Inconel X-750 Sheet, STA	Flex(d)	−1.0	30-40	400	58	455	66	525	76
	Flex(e)	−1.0	30-40	495	72	580	84	705	102
Inconel 718 bar, STA	Axial	0	28	760	110	965	140	1075	156

(a) STA for K Monel: 815 °C (1500 °F), WQ, 590 °C (1100 °F) 16 h plus controlling cooling. STA for Inconel X-750: 980 °C (1800 °F), FC to 700 °C (1300 °F), hold at 700 °C (1300 °F) 20 h, AC. STA for Inconel 718: 1060 °C (1950 °F), AC, 760 °C (1400 °F) 10 h, FC to 650 °C (1200 °F), hold at 650 °C (1200 °F) 10 h, AC. (b) Surface finish, 90 μin. rms. (c) Surface finish, 16 μin. rms. (d) Surface finish, 64 μin. rms. (e) Surface finish, 11 μin. rms.

Chapter 6

Microstructural Degradation, Overheating, Stability

INTRODUCTION

Many superalloys respond to heat treatment, and thus exposure of these alloys to elevated temperatures, with or without stress, can cause microstructural changes that affect properties. Generally, the higher the exposure temperature, the more rapid the structural change. As exposure temperature decreases, the type of microstructural degradation may change. At the highest exposure temperatures, an alloy may be subjected to incipient melting. In addition, oxidation and surface corrosion may occur at temperatures for which these alloys normally are specified.

This chapter deals principally with the effects of microstructural changes, melting, and corrosion on nickel-base and cobalt-base superalloys at temperatures above about 725 °C (1340 °F). It also touches briefly on microstructural changes affecting nickel-base, iron-base, and iron-nickel superalloys at temperatures below 700 °C (1300 °F).

When times or temperatures exceed normal test or operating levels, alloys are exposed to substantially different operating environments from those ordinarily experienced. They may behave in a manner that could not have been predicted from normal test data. Additional information on behavior at elevated temperatures may be found in failure analysis studies.

OVERHEATING

Overheating in the broad sense consists of exposing a metal to excessively high temperatures for short periods of time. Allowable metal temperatures for wrought superalloys in structural applications generally do not exceed about 950 °C (1740 °F). In applications where the component does not bear a load, allowable temperatures may exceed 1200 °C (2200 °F). In general, any temperature can be considered to be in the overheating range when it: (*a*) causes melting; (*b*) causes strengthening phases to dissolve in the matrix or to excessively coalesce; or (*c*) causes extensive oxidation or corrosion. Results of overheating depend on the maximum temperature reached by the metal.

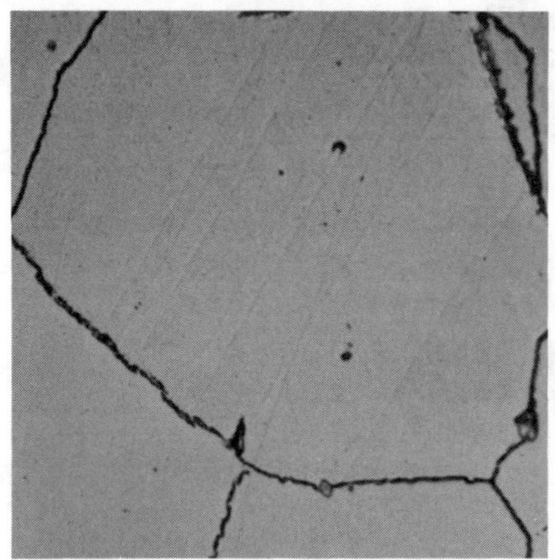

Inconel 617: No melting (500×, unetched)

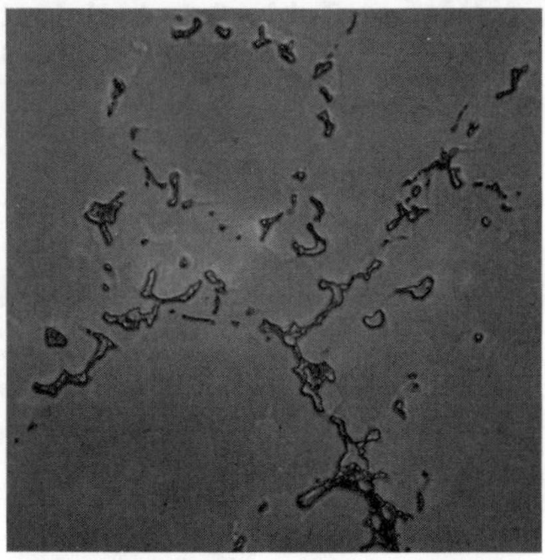

Inconel 617: Incipient melting (500×, unetched)

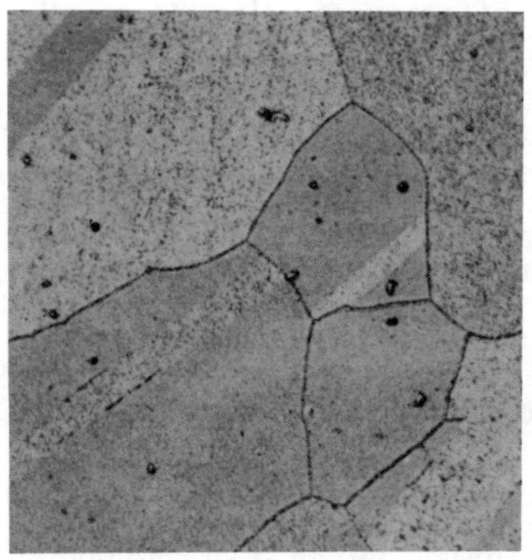

Udimet 700: No melting (500×, lactic acid + HCl + HNO_3 etch)

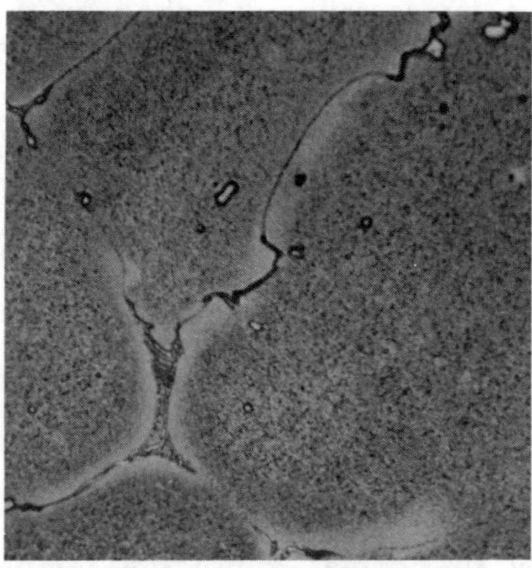

Udimet 700: Incipient melting (500×, lactic acid + HCl + HNO_3 etch)

Fig. 6.1 Effect of incipient melting on microstructure of two nickel-base superalloys

Nickel-base and cobalt-base superalloys generally have incipient melting temperatures above 1200 °C (2200 °F). Table 6.1 gives melting temperatures for some representative wrought superalloys. Figure 6.1 shows the microstructures of two typical wrought superalloys before and after incipient melting. Incipient melting reduces grain-boundary strength and ductility. It thus significantly reduces alloy rupture capabilities.

Once an alloy has exceeded its incipient melting point, normal properties cannot be restored by any known heat treatment. In cast alloys, incipient melting may occur at temperatures substantially below the temperatures predicted from alloy composition. This behavior results from alloy segregation to grain boundaries and

Table 6.1 Incipient melting temperatures of select wrought superalloys

Alloy	Incipient melting temperature °C	°F
Hastelloy X	1250	2280
Haynes 25 (L-605)	1329	2425
Haynes 188	1302	2375
Incoloy 800	1357	2475
Incoloy 825	1370	2500
Inconel 617	1333	2430
Inconel 625	1288	2350
Inconel X-750	1393	2540
Nimonic 80A	1360	2480
Nimonic 90	1310	2390
Nimonic 105	1290	2354
René 41	1232	2250
Udimet 500	1260	2300
Udimet 700	1216	2220
Waspaloy	1329	2425

interdendritic areas during solidification. This is particularly troublesome with cast superalloys, because, based on their excellent high-temperature strength, they often are used for very high-temperature applications, such as jet engine turbine airfoils. In wrought alloys, incipient melting takes place at a temperature much closer to the general alloy melting temperature (solidus), and overheating actually may cause significant portions of the structure to melt.

Effects on Oxidation Resistance

Overheating may deplete alloying elements that provide oxidation resistance. Oxidation attack is thus accelerated on return of the alloy to normal operating temperatures. Even if prolonged exposure to overtemperature does not result in mechanical failure, it frequently will cause excessive surface corrosion. This, in turn, can reduce the strength of the alloy. Even if an alloy is coated for oxidation resistance, the coating will be degraded by surface attack and eventually mechanical properties of the alloy will be affected. The alloy under a coating may also degrade rapidly because of excessive interdiffusion of alloying elements, if temperatures exceed the normal operating range. Coatings eventually fail because of diffusion in both directions - the coating composition diffusing into the substrate and the substrate alloy diffusing outward into the coating. This diffusion changes the coating composition over time so that it no longer provides effective protection. Overheating thus hastens coating failure.

As a rule of thumb, alloys used in structural applications should not be exposed to temperatures within about 125 °C (225 °F) of their incipient melting temperatures. The strength and oxidation resistance of the alloy and the operating environment will determine how close actual metal temperatures may approach the suggested upper limit.

Table 6.2 Solution treatments for select wrought nickel-base superalloys

Alloy	Solution temperature(a) °C	°F	Time, h
Inconel X-750	1150	2100	4
Nimonic 90	1080	1975	8
Nimonic 105	1125-1150	2060-2100	4
Udimet 500	1175	2150	2
Udimet 700	1175	2150	4
Waspaloy	1080	1975	4

(a) All materials air cooled after solution treatment.

Effects on Strength

An important factor to consider when dealing with coated superalloys is the reduction in incipient melting temperature of the system (coating/base metal) that may result from the change in composition brought about by diffusion. For example, for aluminide coatings on an alloy such as U-700, which has an incipient melting temperature of about 1215 °C (2220 °F), incipient melting may occur in the inner diffusion zone at temperatures between 1175 and 1190 °C (2150 and 2175 °F). Incipient melting in coated systems leads to accelerated degradation of the coating.

At temperatures that cause neither incipient melting nor surface degradation, alloy strength may still be reduced because strengthening phases are taken into solution, or they may become less effective as strengtheners because of coalescence. Wrought nickel-base alloys frequently are strengthened by a precipitation of the phase γ', $Ni_3(Al,Ti)$, a face-centered cubic (fcc) intermetallic compound. In wrought alloys, the γ' can be taken into solution at temperatures of about 1175 °C (2150 °F) or less. (See Table 6.2.) Exposure at or near the solution temperature will reduce the amount of γ' by solid solution and thereby reduce alloy strength. Exposure at temperatures below the solution temperature, but still above normal operating temperature, will also reduce strength as the precipitated γ' strengthening particles coalesce to coarser size and become less effective in strengthening.

Prolonged operation at temperatures within the solution range is inadvisable, although occasionally excursions into this range may sometimes be tolerated if the part can be re-heat-treated before excessive creep occurs. If γ' is dissolved, it can be reprecipitated as fine particles by subsequent aging; original property levels can be reasonably recovered, assuming additional damage due to stress or oxidation does not occur. However, properties will not be recovered if slow cooling is used after extensive solution has occurred, or if the material is held at a high temperature so that coarse γ' forms while fine γ' dissolves.

Figure 6.2 compares the service life of a nickel-base superalloy when it is solution treated only and also when it is fully heat treated. Under high-stress applications, strength is significantly reduced if a γ'-strengthened alloy has been exposed to a solution treatment

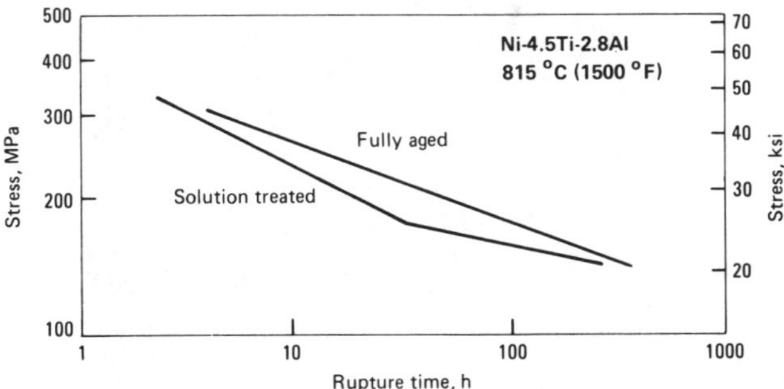

Fig. 6.2 Stress-rupture plot for a γ'-strengthening nickel alloy

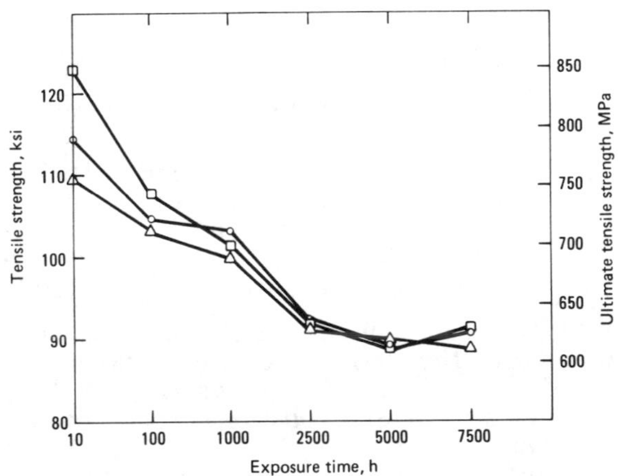

Fig. 6.3 Room-temperature tensile strength vs exposure time at 1040 °C (1900 °F) in air for Hastelloy X

temperature and not re-aged. During long-time, low-stress applications within the aging temperature range, a lesser reduction in strength occurs.

Carbide Phases

Carbide phases in superalloys behave somewhat like γ', but subsequent reprecipitation is not as easily controlled. Carbides are taken into solution or agglomerated during over-temperature exposure, and there can be substantial variations in the amount, form, and distribution of the resulting carbide structure. In Hastelloy X, a solid-solution-strengthened nickel superalloy, there are large differences in the volume fractions and structure of carbides between normal-temperature and overtemperature exposures. The volume fraction of M_6C is about 12 vol% after approximately 7500 h at a temperature of 980 °C (1800 °F). During overtemperature exposure,

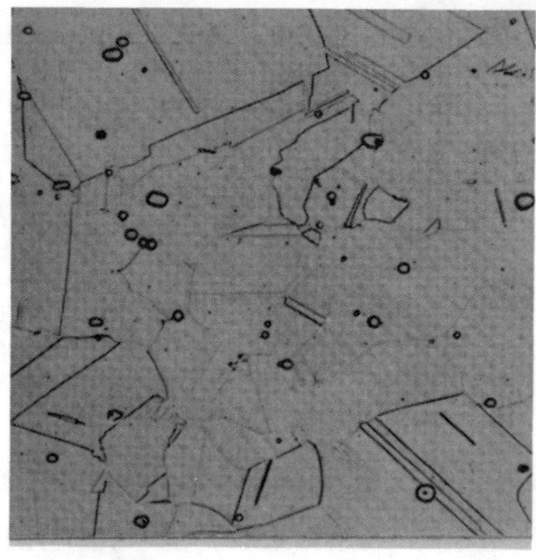

Solution treated

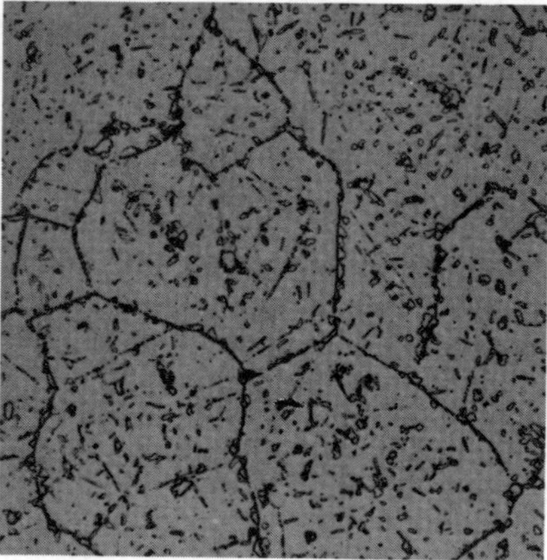

After 6244 h at 870 °C (1600 °F)

500×; etched in HCl + H_2O_2.

Fig. 6.4 Microstructure of Haynes 188

carbide form and distribution are changed considerably due to agglomeration and rounding particles. As a result, mechanical properties are reduced, as shown in Fig. 6.3, for room-temperature strength after exposure at 1040 °C (1900 °F) in air. Similar results were obtained from a series of exposures at about 1100 °C (2000 °F).

Extensive carbide precipitation frequently can occur in alloys when there is a change from the original carbide that was present in the mill annealed condition. Figure 6.4 shows Haynes 188, a cobalt-base wrought alloy, before and after exposure for approximately 6000 h at 870 °C (1600 °F). During exposure, M_6C carbides dissolved and were replaced predominantly by $M_{23}C_6$ carbides and, to a lesser extent, by Laves phase. Significant losses in ductility resulted from the precipitation shown in Fig. 6.4.

In alloys where extensive carbide precipitation or agglomeration occurs, it is frequently possible to recover the original carbide distribution, and a major portion of alloy properties by solution heat treatment. For example, it is claimed that Haynes 188 can recover original room-temperature ductility by heat treatment at 1175 °C (2150 °F) for 15 min.

ALLOY STABILITY

Obviously, alloy stability and the ability of a material to resist damaging effects due to overheating are of particular importance to superalloys that are intended for elevated temperature service. Clearly, the nickel-base superalloys strengthened by a secondary precipitated phase are the most complex and indeed the most remarkable of all the superalloys. These alloys are used in the most demanding applications relative to stress and temperature. They have demonstrated remarkably useful strength at the highest fraction of the base metal melting point of any alloy system ever developed.

As discussed above, a significant characteristic of high-temperature service is metallurgical instability. Stress, time, temperature, and environment may act to change the metallurgical structure during operation and, thereby, contribute to failure by reducing strength and/or ductility. It should be noted that in a few cases strength may be enhanced. These structural changes or metallurgical instabilities are best described in terms of their influence on stress-rupture properties. A sharp change downward in the slope of the stress-rupture curve indicates that failures will occur in shorter times and at lower stresses than originally predicted. Instabilities are associated with aging (phase precipitation), overaging (phase coalescence and coarsening), phase decomposition (usually carbides, borides, and nitrides), intermetallic phase precipitation, internal oxidation, and stress-corrosion.

Typical instability problems with the γ' nickel-base superalloys involve intermetallic phase formation. In certain alloys where composition has not been carefully controlled, undesirable hard, brittle phases can form either during heat treatment or service. These hard compounds have been identified as close-packed σ, μ, and Laves phases, and it has been determined that they form at elevated temperatures with a deteriorating effect on stress-rupture properties. These phases are characterized by platelike structures, often nucleating on grain-boundary carbides.

Still another instability characteristic relates to carbide reactions. A variety of carbides are found in superalloys identified by this chemical composition, such as TiC, Cr_7C_3, etc. In many instances, the carbide composition is complex and is comprised of many of the superalloy alloying elements, such as chromium, molybdenum, tungsten, niobium, and tantalum. These complex metal combinations are identified by the letter "M". Thus, superalloys are characterized by carbides identified as MC, M_6C, and $M_{23}C_6$. Although temperature and stress affect both carbides within grains and at grain boundaries, the effects on grain-boundary carbides are usually a much more significant factor in altering creep and rupture behavior. Grain-boundary morphology is indeed important relative to high-temperature properties. The presence of carbides at grain boundaries as strengtheners is necessary for optimum creep and rupture life, but alteration in shape or breakdown to other carbide forms may cause property degradation. The best carbide formation at grain boundaries for optimum strength is discrete, blocky particles. Continuous carbide films at the boundaries substantially reduce stress-rupture life.

In superalloy selection, it is important to ensure that the composition is such that carbide morphology at grain boundaries is proper and that detrimental σ, μ, or Laves phases do not form at operating temperatures.

MICROSTRUCTURAL DEGRADATION

Effects of Service Below 700 °C (1300 °F)

For superalloys normally used at temperatures of 425 to 725 °C (800 to 1340 °F), melting by overtemperature is not a problem;

microstructural changes are the important consideration. Hot corrosion and other types of accelerated oxidation are important for some alloys. The superalloys used at these temperatures are relatively stable. Superalloys such as A-286, Incoloy 901, V-57, Waspaloy, and Astroloy generally are considered microstructurally stable at the temperatures for which they are used as disks. Their properties are almost exclusively determined by prior heat treatment.

If nickel-base alloys such as Waspaloy, Astroloy, and Rene′ 95 are exposed for prolonged times at metal temperatures that exceed about 650 °C (1200 °F), the strengthening phase may coarsen slightly. The same effect may occur above 600 °C (1100 °F) for iron-base or iron-nickel superalloys such as A-288. However, unless the alloys are operated at temperatures well into or above their aging-temperature ranges, no significant microstructural effects will be noted. Carbide precipitation at dislocations may be significant when operating times at 480 to 650 °C (900 to 1200 °F) exceed 10,000 h. However, there are no published data to support the existence of significant effects of thermal exposure on creep-rupture behavior, although notch behavior may become important. At these temperatures, surface oxidation generally is not a problem, although long-term exposure to certain highly corrosive environments may produce surface attack such as hot corrosion (sulfidation).

Inconel 718 deserves some mention because it often is used at temperatures above its final aging temperature of 620 °C (1150 °F). Although some minor coarsening of the γ'' phase may take place, no detrimental effects normally occur at temperatures up to 650 °C (1200 °F). If overheating to above 700 °C (1300 °F) should occur, the strengthening γ'' phase is degraded; significant σ phase precipitation can then occur, with resultant losses in strength. With respect to carbide changes, superalloys are not as sensitive to corrosive attack as some austenitic stainless steels are because of the many carbide formers in superalloy compositions.

Effects of Service Above 700 °C (1300 °F)

Whereas some superalloys are exposed to extremely high temperatures in furnace and petrochemical applications, alloys for gas turbine applications generally are exposed to the most demanding combinations of high temperature and stress. The normal operating regime of turbine blades and vanes is about 725 to 850 °C (1340 to 1560 °F) for wrought alloys and up to 1050 °C (1920 °F) for cast alloys. Within these ranges, microstructural changes readily occur with time at temperature. Furthermore, when stress is applied, the changes may be accelerated.

The principal changes are: (*a*) breakdown of primary carbides and formation of secondary carbides; (*b*) agglomeration of primary geometrically close-packed (gcp) strengthening phases such as γ'; and (*c*) formation of topologically close-packed (tcp) phases, such as σ, Laves, and μ. Processes described under (*a*) and (*b*) are an extension of the normal strengthening process. These reactions are recognized during the design of components and allowances are made for their effects on alloy strength at moderate times. Carbide transition and γ' agglomeration do reduce strength with time, but generally are not as detrimental as formation of tcp phases.

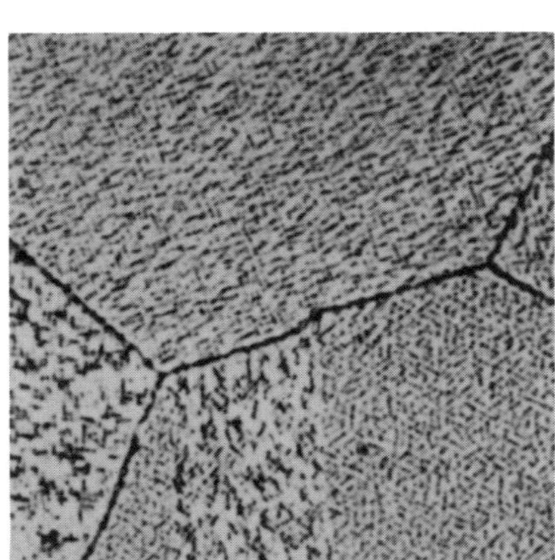

Solution treated and aged

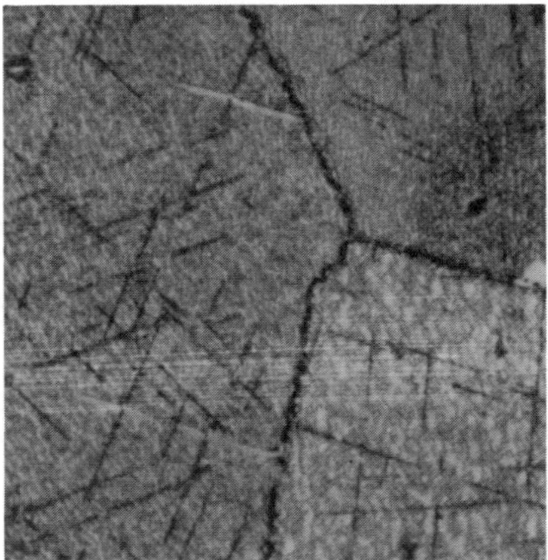

After 500 h at 870 °C (1600 °F)

100×; etched in Kalling's reagent.

Fig. 6.5 Microstructure of Udimet 700

Figure 6.5 shows the microstructure of Udimet 700 nickel-base alloy before and after exposure at 870 °C (1600 °F). During exposure, the γ' became agglomerated and $M_{23}C_6$ carbides formed. Generally, design parameters take into account γ' agglomeration if it is experienced in the moderate times used to test alloys (most often, 20 to 1000 h). However, for significantly longer times at normal temperatures, γ' agglomeration may reduce alloy strength below predicted values.

Changes in carbide phases also adversely affect strength, although initially there may be increases in strength as additional carbides precipitate. In the wrought cobalt-base solid-solution alloy Haynes 25 (L-605), carbide precipitation at 815 °C (1500 °F) is responsible for alloy hardening in both early and late stages of exposure. In the late stages, $M_{23}C_6$, M_6C, and Laves phase participate in strengthening. In Hastelloy X, extensive precipitation occurs at 705 to 790 °C (1300 to 1450 °F). As a result, σ, μ, and a dense intragranular secondary M_6C carbide can occupy as much as 27 vol% of the structure. As primary M_6C carbides coalesce, there is a continual reduction in strength; formation of small secondary carbides and tcp phases enhances strength, but also reduces ductility.

Chapter 7

Metallography

INTRODUCTION

As discussed previously (Chapters 3 and 6), superalloys encompass a wide range of microstructural constituents and phases. This chapter provides a brief summary of the metallographic techniques required to study superalloy microstructures and to identify their respective phases and constituents. For more detailed information, see "Wrought Heat-Resistant Alloys" in Volume 9 of the 9th Edition of *Metals Handbook*.

SPECIMEN PREPARATION

In general, the procedures used to prepare metallographic specimens of superalloys are quite similar to those used for most metals. Aspects particularly significant to the metallographic specimen preparation for superalloys will be emphasized. Nominal compositions of the superalloys discussed in this chapter are provided in Table 2.1, Chapter 2. There are two general types of metallographic inspection:

* Macroexamination
* Microexamination

Macroexamination, or macroetching, is differentiated from microexamination mainly by the magnification used. Macroexamination is accomplished with minor magnification at best, whereas microexamination involves significant enlargement of the specimen, as much as 50,000 times the actual size. Considerable microexamination, however, is done at 100 to 500 times the actual size. Furthermore, macroexamination is used for "rough eye" evaluation as compared to the precise study of microstructure afforded by microexamination, which uses sophisticated microscopes.

MACROEXAMINATION

To macroetch wrought materials, disks are cut (usually from billet samples) at representative locations, such as the top and bottom of ingots or remelted stock, and are ground before macroetching.

Table 7.1 Macroetchants for superalloys

Composition	Comments
Composition 1	
1 part 30% H_2O_2, 2 parts HCl, 2 or 3 parts H_2O	Recommended for nickel-chromium alloys; use fresh solution; reveals grain structure; if the surface is stained, remove with 50% aqueous HNO_3; etch approximately 2 min
Composition 2	
(a) 15 g $(NH_4)_2S_2O_8$ (ammonium persulfate) and 75 mL H_2O (b) 250 g $FeCl_3$ and 100 mL HCl (c) 30 mL HNO_3	Macroetch for general macrostructure and weldments; mix (a) and (b), add (c); immerse 30-120 s at room temperature
Composition 3	
200 g $FeCl_3$, 200 mL HCl, H_2O to 1000 mL	For nickel-base superalloys; etch to 90 min at 100 °C (212 °F)
Composition 4	
HCl saturated with $FeCl_3$	For cobalt-base superalloys; add 5% HNO_3 before use at room temperature; clean surface by dipping in 50% aqueous HCl
Composition 5	
(a) 21 mL H_2SO_4, 15 mL HCl, 21 mL HNO_3, 21 mL HF, 22 mL H_2O (b) 40 mL 20% aqueous $CuCl_2$ (copper chloride), 40 mL HCl, 20 mL HF	For cobalt-base superalloys; etch 5 min in (a), then 5 min in (b)
Composition 6	
50 mL saturated aqueous $CuSO_4$ (copper sulfate) and 50 mL HCl	For iron-nickel- and nickel-base alloys; swab or immerse, room temperature

Note: Whenever water is specified, use distilled water.

Macroetchants used for wrought austenitic iron-nickel-chromium superalloys are the same as those recommended for wrought austenitic stainless steels. Macroetchants for various superalloys are given in Table 7.1.

Macroetched features in consumable-electrode remelted superalloys are different from those observed in ingot cast steels. Unique macroetch features observed in these remelted alloys include freckles, radial segregation, and ring patterns. Freckles are dark-etching spots caused by localized segregation, or by enrichment of carbides or Laves phase. They are detrimental to material quality. Radial segregation appears as dark-etching elongated spots in a radial or spiral pattern. Ring patterns are concentric rings that etch lighter (usually) or darker than the matrix. They are revealed only by macroetching. The nature of ring patterns remains obscure. Examination has revealed little difference between ring and matrix areas and no measurable influence on mechanical properties.

Superalloys are examined for macrostructure in the same manner as tool steels and stainless steels. Macroexamination is associated primarily with wrought superalloys, although castings can be macroexamined by approximate grain size.

MICROEXAMINATION

Sectioning

Various sectioning devices have been used with superalloys. The usual precautions regarding excessive heating should be followed. Because superalloys are sensitive to deformation, the more vigorous sectioning techniques, such as band sawing or power hacksawing, may introduce excessive distortion or work hardening, depending on the alloy and heat treated condition. Such methods are suitable for initial sectioning of large pieces, but final cutting of specimens is best performed using abrasive cutoff machines. Heavy deformation introduced by sawing may be difficult to remove by subsequent grindings and polishings. Typical abrasive cutoff wheels usually are of the consumable type. Abundant coolant should be used, and submerged cutting is preferred.

Mounting

Many bulk specimens can be ground and polished without mounting. Some automatic grinding and polishing devices require a mounted specimen, which is usually 25, 32, or 38 mm (1, 1$^1/_4$, or 1$^1/_2$ in.) in diameter; others do not. Mounting facilitates polishing of small or irregularly shaped specimens.

Sometimes, it may be necessary to examine the microstructure at the extreme edge of a specimen. Optimum results are obtained when the edge of interest is plated with electroless or electrolytic nickel before mounting. Good results can be obtained without plating, if the specimen is mounted using a compression-molded epoxy resin, particularly if automatic polishing is used.

When edge retention is not a primary requirement, specimens can be mounted using any of the popular compression-mounting materials or castable resins. Most produce acceptable results, although each resin has advantages and disadvantages. Selection of a particular resin frequently is based on familiarity.

Grinding

Specimens may be ground by hand, or by automatic devices. Grinding is performed using water-cooled silicon carbide paper at 150 to 600 rpm. The typical grit size sequence is 120, 240, 320, 400, and 600 grit. Finer grits occasionally are used. Grit sizes finer than 600 can be useful for preparing nickel-base superalloys. This practice eliminates coarse diamond paste polishing and may improve retention of carbides and intermetallic phases, as well as reduce relief.

Moderately heavy pressure should be used for hand grinding. The specimen must be held flat against the paper. After each grinding, the specimen should be rinsed, wiped clean, and rotated 45 to 90° before grinding with the next paper. Grinding should proceed for approximately twice as long as needed to remove all scratches from the previous step; 1 to 2 min per step is usually adequate. Automatic grinding produces omnidirectional scratch patterns. Wrought superalloys are not susceptible to embedding of silicon carbide from the grinding paper, but if the specimen contains cracks or pores, it should be cleaned ultrasonically. For most specimens, washing under running water will remove any loose abrasive or grinding debris. Specimens should be washed in alcohol (preferably ethanol) and dried in hot air.

Rough Polishing

Often rough polishing begins using 6- or 3-μm diamond abrasive, generally as a paste, although aerosols or slurries also are used. The use of 5-μm alumina (Al_2O_3) for rough polishing has been replaced primarily by the more efficient diamond abrasives.

Although a wide range of cloths are available, low-nap or napless cloths generally are used for rough polishing. Canvas, which is quite popular, provides economical durability. Synthetic napless chemotextile cloths also are used. Billiard cloth and red felt are sometimes preferred. A lubricant/extender fluid should always be added to reduce friction and drag and to promote more efficient cutting. Wheel speeds of 150 to 300 rpm are suggested. Moderate pressure should be applied.

During hand polishing, the specimen should be rotated in a direction counter to wheel rotation while it is moved slowly from center to edge. Again, the specimen must be held firmly against the wheel to avoid rocking. Polishing should continue until grinding scratches are removed; 1 to 2 min is usually adequate.

A second diamond polishing step using 1-μm diamond is often performed. A synthetic suede medium-nap cloth is commonly used, but other cloths also may provide good results. This step is carried out in the same manner as the initial diamond polishing. After each diamond polishing, the specimen should be cleaned carefully to remove abrasive, extender oil, and polishing debris. Ultrasonic cleaning produces excellent results, but is not always required.

Final Polishing

One or more steps may be involved in final polishing, depending on the need to remove all scratches. For routine inspection, polishing to a 1-μm diamond finish may be adequate. When photomicroscopy is anticipated, additional polishing steps usually are required. A wide range of final polishing abrasives may be used. Alumina slurries containing 0.3-μm and/or 0.05-μm Al_2O_3 are used frequently. Colloidal silica (SiO_2) produces exceptional results with these alloys.

Final polishing generally is performed using synthetic suede medium-nap cloths and wheel speeds of approximately 150 rpm. During hand polishing, the specimen should be rotated in a direction counter to wheel rotation while it is moved from center to edge. Slurry occasionally is added to the cloth during polishing. Moderate pressure should be used, and care must be taken to avoid rocking the specimen. A polishing time of 1 to 2 min is usually adequate.

Table 7.2 Electropolishing techniques for wrought superalloys

Electrolyte composition	Comments
1. 37 mL H_3PO_4, 56 mL glycerol, 7 mL H_2O	For Inconel 625, use 1.2–1.6 A/cm² (8–10 A/in.²); for Incoloy 800, use 3.1 A/cm² (20 A/in.²); platinum cathode
2. 25 mL H_3PO_4, 25 mL HNO_3, 50 mL H_2O	For Inconel 600 and X-750, use 17.8 A/cm² (115 A/in.²), 5–10 s; platinum cathode
3. 144 mL ethanol, 10 g $AlCl_3$ (anhydrous aluminum chloride), 45 g $ZnCl_2$ (anhydrous zinc chloride), 16 mL N-butyl alcohol, 32 mL H_2O	For cobalt-base superalloys; use 23–25 V dc at room temperature with successive 1-min periods
4. 40 mL $HClO_4$ (perchloric acid), 450 mL acetic acid, 15 mL H_2O	For Nimonic alloys; use 15 V dc, 0.1 A/cm² (0.65 A/in.²) below 25°C (77°F)
5. 10 mL $HClO_4$ and 90 mL acetic acid	For nickel-base alloys, use at 0.5–0.9 A/cm² (3–6 A/in.²), 30 s for aged specimens, longer for solution-annealed ones; keep cool (10–15 °C, or 50–60 °F); best results by polishing in 5-s intervals
6. 70 mL methanol and 10 mL H_2SO_4	For nickel-base superalloys, use at 20–25 V dc, 0.3–0.8 A/cm² (2–5 A/in.²), room temperature; 10–15 s after a 600-grit finish or 5 s after a 0.3-μm Al_2O_3 finish; γ' slightly etched, carbides in relief
7. 60 mL methanol, 10 mL H_2SO_4, 5 mL HCl	Use same as No. 6; produces more etching of γ' phase; if HNO_3 is substituted for HCl, the surface will be smooth without relief or attack
8. 7 mL ethanol, 20 mL $HClO_4$, 10 mL glycerol	Use same as No. 6; mix carefully, keep cool; produces smooth surfaces
9. 15 mL HCl and 85 mL methanol	For nickel-base superalloys, use at 30-40 V dc, 0.3–1.2 A/cm² (2–8 A/in.²), at room temperature for 5–10 s after a 600-grit silicon carbide finish or 2–5 s after a 0.3-μm Al_2O_3 finish; produces strong carbide relief, etches γ'; very good for SEM examination

After polishing, the specimen should be cleaned carefully and dried to avoid staining. When hand polishing, the operator's hands must be cleaned thoroughly after each rough and final polishing to prevent contamination.

Automatic polishing machines, a number of which are currently available, are quite useful for final polishing. The time required using these units ranges from a few minutes to approximately 30 min for vibratory polishing.

Mechanical polishing is sometimes followed with a brief electropolish to remove any smeared or flawed metal, without introducing preferential attack of the second phase constituents. Extended electropolishing should be avoided. Table 7.2 lists appropriate electropolishing solutions for various superalloys. Electropolishing solutions for wrought iron-nickel-cobalt superalloys are the same as those used for wrought austenitic stainless steels.

Etching

Some minor phases in superalloys can be observed easily in the as-polished condition. Minimal relief can be introduced during final polishing to accentuate these particles. They can be observed in bright-field illumination by color contrast, which is useful for phase identification. Nitrides, carbonitrides, borides, and metal carbides can be observed readily without etching.

Wrought iron-nickel-base superalloys are metallographically similar to austenitic stainless steels. Therefore, the techniques for etching and identifying phases in wrought austenitic stainless steels apply to these alloys.

Glyceregia is one of the most prevalent reagents for revealing the general structure of iron-nickel and nickel-base superalloys. It should always be mixed fresh and discarded when it turns orange. Glyceregia is not effective with some of the high-cobalt superalloys. For etching solution-annealed nickel-base alloys, glycerol content frequently is decreased, whereas nitric acid (HNO_3) content is often increased. The standard composition is ideal for solution-annealed and aged specimens, which are easier to etch. The mixture of hydrochloric acid, sulfuric acid, and nitric acid (HCl-H_2SO_4-HNO_3) (95-5-3) is also quite popular for these alloys and is used similarly.

For grain size examination of aged specimens, etching with waterless Kalling's or Marble's reagent is quite common. Several electrolytic reagents are also commonly used. Color etchants, although not widely used for these alloys, produce very good results. Table 7.3 lists some of the more commonly used reagents for etching iron-nickel, nickel, and cobalt superalloys.

Some electropolishing procedures are useful for microstructural examination using scanning electron microscopy (SEM) or transmission electron microscopy (TEM) with replicas. The electropolish sometimes produces attack of the γ' phase, and subsequent etching is unnecessary. For replica examination, it is often helpful to use a reagent or an electropolish that attacks γ' so that this phase is readily distinguished from other phases. In such cases, the γ' is recessed, but other second phases are in relief.

Selective etchants and heat tinting commonly have been used to differentiate various carbide types and to identify σ phase. Borides, which are similar in appearance to metal carbides, can be discriminated by selective etching. Metal carbides are colored selectively, whereas borides are unaffected.

MICROEXAMINATION TECHNIQUES

Transmission Electron Microscopy

Because some of the important constituents in superalloys, such as γ' phase, are generally too small to observe using the optical microscope, considerable use has been made of transmission electron microscopy (TEM). In addition to affording greater resolution and higher magnification, TEM provides a means for phase identification by electron diffraction and, when equipped with X-ray detectors, can provide chemical analysis data. Such analytical procedures are necessary to understand the strengthening mechanisms for superalloys.

Several types of specimens can be prepared for TEM examination; each type has advantages and disadvantages. The replica method, which had been used predominantly, is being replaced by use of the scanning electron microscope. A well-polished and properly etched specimen can be examined by scanning electron microscopy (SEM) at magnifications of 50,000× or more. Therefore, much of the structural examination role of TEM replicas can be accomplished without replica preparation and the complication of replica interpretation or artifact control.

Table 7.3 Microetchants for superalloys

Composition	Comments
1. 3 parts glycerol, 2-3 parts HCl, 1 part HNO_3	Glyceregia; mix fresh, do not store; discard when solution is orange; use by immersion or swabbing 5-60 s; very popular general etch for structure of iron-base and nickel-base superalloys; γ' in relief
2. 5 mL HF, 10 mL glycerol, 85 mL ethanol	Electrolytic etch at 0.04-0.15 A/cm² (0.25-1.0 A/in.²), 6-12 V dc; for nickel-base alloys, γ' in relief; stop etch when edges are brownish; excellent etch for TEM replica work
3. 12 mL H_3PO_4, 47 mL H_2SO_4, 41 mL HNO_3	Electrolytic etch, 6 V dc, 0.12-0.15 A/cm² (0.75-1.0 A/in.²), a few seconds; add to 100 mL H_2O to slow etch; for nickel-base alloys; use under hood; mix H_3PO_4 and HNO_3, then add H_2SO_4; stains matrix when γ' is present; good for revealing segregation and for examining γ' with TEM replicas; attacks Bakelite; stop etch when edge of specimen is brownish
4. 30 mL lactic acid, 20 mL HCl, 10 mL HNO_3	For nickel-base superalloys
5. 5 g $CuCl_2$ 100 mL HCl, 100 mL ethanol	Waterless Kalling's reagent; immerse or swab to a few minutes; for iron-base and nickel-base superalloys
6. 10 g $CuSO_4$, 50 mL HCl, 50 mL H_2O(a)	Marble's reagent for iron-nickel and cobalt-base superalloys; immerse or swab 5-60 s; a few drops of H_2SO_4 will increase etch activity; reveals grain boundaries and second-phase particles
7. 5 mL H_2SO_4, 3 mL HNO_3, 92 mL HCl	For iron-base and nickel-base alloys; add H_2SO_4 to HCl, stir, allow to cool, add HNO_3; discard when orange; swab 10-30 s; use under hood; do not store
8. 20 mL HNO_3 and 60 mL HCl	Aqua regia; for iron-base and nickel-base superalloys; use under hood, do not store; immerse or swab 5-60 s; attacks σ phase, outlines carbides, reveals grain boundaries
9. 50 mL HCl and 1-2 mL 30% H_2O_2	For nickel-base alloys; attacks γ'-phase; immerse 10-15 s
10. 5 mL H_2SO_4 8 g CrO_3, 85 mL H_3PO_4	Electrolytic etch at 10 V dc, 0.2 A/cm² (1.3 A/in.²), 5-30 s; reveals inhomogeneities in nickel-base alloys
11. 10 mL HNO_3, 10 mL acetic acid, 15 mL HCl, 2-5 drops glycerol	Use fresh, same precautions as glyceregia; used for hard to etch solution-treated nickel-base alloys
12. (a) 33 mL HCl and 67 mL H_2O (b) 50 mL HCl and 50 mL H_2O	Beraha's tint etch for nickel-base alloys; add 0.6-1 g $K_2S_2O_5$ (potassium metabisulfite) to 100 mL stock solution (a); immerse (never swab) 60-150 s, slowly agitate; if colors are not developed, add 1-1.5 g $FeCl_3$ or 2-10 g $NH_4F \cdot HF$ (ammonium bifluoride) to 100 mL stock solution (b); immerse 60-150 s, agitate gently; colors matrix
13. 10 g $K_3Fe(CN)_6$ (potassium ferricyanide), 10 g KOH, 100 mL H_2O	Murakami's reagent; for iron-base and nickel-base superalloys; use hot (75 °C, or 170 °F) to darken σ phase; use at room temperature to darken carbides; better results may be obtained if the specimen is first etched in 50% aqueous HNO_3 at 8 V dc; use under a hood
14. 10 g CrO_3 and 100 mL H_2O	Electrolytic etch, 6 V dc, 10-30 s; for iron-base and nickel-base superalloys; σ attacked, carbides outlined or attacked
15. 80 mL H_3PO_4 and 10 mL H_2O	Electrolytic etch for nickel-base superalloys at 3 V dc (closed-circuit), 0.11-0.12 A/cm² (0.7-0.8 A/in.²), 7-9 s; if the surface is stained, swab with the electrolyte; use fresh solution; used to determine the degree of carbide continuity at the grain boundaries
16. 25 g CrO_3, 130 mL acetic acid, 7 mL H_2O	Electrolytic etch for nickel-base superalloys at 10 V dc (closed circuit) for 2 min; the current density will drop during the first 20 s; use fresh; used to reveal prior grain boundaries
17. 30 mL HCl, 7 mL H_2O, 3 mL 30% H_2O_2	Popular etch for cobalt-base superalloys
18. 100 mL HCl and 5 mL 30% H_2O_2	Popular etch for cobalt-base superalloys; up to 20% H_2O_2 has been used; mix fresh; immerse 1-10 s
19. 5-10 mL HCl and 95-90 mL H_2O	Electrolytic etch for cobalt-base superalloys; use at 3 V dc, 1-5 s, carbon cathode
20. 2 mL H_2SO_4 and 98 mL H_2O	Etch first with glyceregia to dissolve matrix uniformly; then etch electrolytically at 6-12 V dc, 0.12-0.15 A/cm² (0.75-1.0 A/in.²) until edge of specimen is brownish; good for TEM replica studies
21. 5 mL HF, 10 mL glycerol, 10-50 mL ethanol, H_2O to 100 mL total volume	Etch first with glyceregia to dissolve matrix uniformly; then etch with solution at left to dissolve γ'; use at 6-12 V dc, 0.12-0.15 A/cm² (0.75-1.0 A/in.²) for less than 1 s; good for TEM replica work or SEM examination

(a) When water is specified, use distilled water

Table 7.4 Electropolishing solutions for TEM thin foils of superalloys

Composition	Comments
1. 950 mL acetic acid and 50 mL $HClO_4$	Popular electropolish for wrought superalloys for perforation; use at 70–80 V dc, 100–120 mA, 15 °C (60 °F)
2. 133 mL acetic acid, 25 g CrO_3, 7 mL H_2O	Best for window method; opacity makes jet thinning difficult; use at 10–12 V dc, 20 °C (70 °F)
3. 77 mL acetic acid and 23 mL $HClO_4$	For cobalt-base superalloys; keep temperature below 30 °C (85 °F), stainless steel cathode; used with the window method; use at 22 V dc, 0.08 A/cm² (0.5 A/in.²)
4. 600 mL methanol, 250 mL butanol, 60 mL $HClO_4$	Two step procedure: (a) 0.13-mm (0.005-in.) disk, polished 15–30 min at 30 V (b) final thinning at 16–24 V; use at −60 to −70 °C (−75 to −95 °F)

In addition, the contrast mechanisms utilized in the scanning electron microscope are valuable for structural examination. Because chemical analysis using SEM is limited to features larger than a few microns, TEM examination and analysis of extracted constituents remains an important procedure. Procedures for preparing structural and extraction replicas are discussed in the article "Transmission Electron Microscopy," Volume 9 of the 9th Edition of *Metals Handbook*. Direct examination of the fine structure of superalloys is also performed by TEM examination of thin foils. As with extraction replicas, electron diffraction and chemical analysis can be performed.

Because the beam size in a transmission electron microscope or a scanning transmission microscope is much smaller than that used in a scanning electron microscope, much finer particles can be analyzed using thin foils without interference from the surrounding matrix. Extremely small particles are difficult to analyze, even with a scanning transmission electron microscope. Extraction replicas are useful, because matrix effects can be eliminated. Additionally, with the use of a scanning transmission electron microscope, microdiffraction patterns can be obtained from individual particles rather than from many particles. The microdiffraction pattern is of great value in basic structural studies of the constituents.

Thin foils are prepared by the window method or the disk method. These methods involve careful sectioning of the specimen to obtain a relatively thin slice of material that is free of artifacts, followed by mechanical, chemical, or electrolytic thinning of the sample until a small area is thin enough for electron transmission. Table 7.4 lists several popular electropolishing procedures for preparing thin foils of superalloys.

Bulk Extraction

X-ray diffraction study of phases that have been extracted electrolytically is a widely used tool for phase identification in superalloys. Because of the complex nature of these alloys, such techniques must be controlled carefully to ensure optimum results. Qualitative identification of phases by this method is considerably easier than quantitative evaluation. The extraction method must be designed to permit separation of the carbides, nitrides, γ', and topologically close-packed (tcp) phases. Once separated, the phases

Table 7.5 Techniques for bulk electrolytic extractions

Solution	Comments
1. 10% HCl in methanol	For extraction of carbides, borides, topologically close-packed phases, and geometrically close-packed phases from nickel-base and iron-nickel-base alloys; solution may dissolve Ni_3Ti; maintain bath at 0–30 °C (32–85 °F); for alloys containing tungsten, niobium, tantalum, or hafnium, add 1% tartaric acid; use 0.05–0.1 A/cm^2 (0.33–0.65 A/in.^{2}) for 4 h or longer (additional details in ASTM E 963)
2. 0.5–2% citric acid and $(NH_4)_2SO_4$ in H_2O (1% of each is most common)	Used to extract γ' phase in nickel-base Udimet 700; use at 0.03 A/cm^2 (0.2 A/in.^{2}) for 3 h at room temperature; minor amounts of carbides and borides will also be extracted
3. 10 or 20% H_3PO_4 in H_2O	Used to extract γ' in nickel-base alloys; may etch the γ' phase, although results have been contradictory
4. 50 mL HNO_3, 20 mL $HClO_4$, 1000 mL H_2O	For extraction of γ' and η in nickel-base superalloys; use at 0.1 A/cm^2 (0.65 A/in.^{2}), 25 °C (75 °F)
5. 300 g KCl (potassium chloride), 30 g citric acid, 50 mL HCl, 1000 mL H_2O	For extraction of carbides from nickel-base superalloys; use at 0.1 A/cm^2 (0.65 A/in.^{2}), 25 °C (75 °F)

can be analyzed using X-ray diffraction, chemical analysis (elemental), and optical and electron microscopy.

Considerable research has been conducted to establish reliable procedures for use of bulk extraction in superalloys. Anodic dissolution using 10% HCl in methanol, which dissolves γ' and the austenitic matrix, is utilized to extract carbides, borides, nitrides, and tcp phases. If the alloy to be digested contains substantial amounts of tungsten, tantalum, or niobium, 1% tartaric acid is added to prevent contamination of the residue.

To extract γ' from nickel-base alloys, two electrolytes have been used: (*a*) 20% aqueous phosphoric acid (H_3PO_4), or (*b*) an aqueous solution containing 1% ammonium sulfate [$(NH_4)_2SO_4$] and 1% citric acid or tartaric acid. The latter electrolyte produces better recovery of γ'. When the ammonium sulfate/citric or tartaric acid electrolyte is used, the residue will also contain carbides, nitrides, and borides (if present in the alloy). All of the γ' morphologies are extracted using this electrolyte. Techniques for bulk electrolytic extraction are given in Table 7.5.

SOLIDIFICATION STRUCTURES IN WELDED JOINTS

Metallographic techniques are used in special cases such as studying the solidification structures in welded superalloy joints. Fusion metal in a welded joint solidifies by epitaxial growth from partially melted grains in the heat-affected zone of the weld. Cells or dendrites grow in packets from each heat-affected zone grain into the weld pool in a preferential crystallographic direction. One consequence of epitaxial growth is that a single, favorably oriented grain (packet) can traverse many successive beads in a multiple-pass weld.

As solidification progresses, grains bend to follow the maximum thermal gradient. This phenomenon can be used to map solidification isotherms through the use of longitudinal, transverse, and normal sections.

During solidification, alloying elements may be rejected by the solid into the liquid regions between the dendrites, cells, and/or neighboring grains. With sufficient segregation, the solidification mode (cellular or dendritic) can be determined using appropriate etches. The pattern of oxides entrapped between cells or dendrites may also reveal the manner in which the weld metal solidified.

It is often difficult to observe the as-solidified structure, because the microstructure is altered by one or more solid-state phase transformations that occur during cooling. However, in some materials, the solidification grain or packet boundaries act as preferential sites for the nucleation of solid-state transformation products. In these cases, the solidification grain structure is marked by discontinuities in the transformation structure.

Solidification structures in the welds of superalloys are usually discernible due to the segregation of one or more solute elements during solidification. For example, in the iron-base A-286 alloy, titanium, phosphorus, and sulfur strongly segregate to the interdendritic regions, allowing the solidification structure to be revealed by a phoschromic electroetch. In aged welds, the solidification structure may be inferred from the inhomogeneous distribution of γ' precipitates.

Nickel-base superalloys with low aluminum and titanium contents are considered to be readily weldable. In these alloys, the solidification structure can be revealed by electroetching with phoschromic or a solution of 10 mL phosphoric acid (H_3PO_4) plus 50 mL sulfuric acid (H_2SO_4) plus 40 mL HNO_3. A technique of etch polishing with a 1:1 solution of acetic acid and HNO_3 followed by etching with glyceregia also produces good results with many nickel-base superalloys. Special care must be taken to avoid smearing the sample, however.

For further information on this subject, see the Section "Metallographic Techniques and Microstructures: Specific Metals and Alloys" in Volume 9 of the 9th Edition of *Metal Handbook*.

Chapter 8

Castings

INTRODUCTION

In terms of tonnage, the most important use of superalloy castings is in metallurgical and other industrial furnaces. Iron-base alloys are used most frequently for this service, although significant amounts of nickel-base and cobalt-base alloys also are used. Other major applications for superalloy castings include turbochargers, gas turbines, power plant equipment, and equipment used in the manufacture of glass, cement, synthetic rubber, chemicals, and petroleum products.

There are two distinct superalloy casting classifications: (*a*) iron-chromium-nickel compositions identified as heat-resistant castings, and (*b*) the nickel-base and cobalt-base superalloys. In the heat-resistant classification, some compositions truly are stainless steels; these alloys will not be discussed in detail in this chapter. Heat-resistant alloys that contain substantial chromium and nickel contents (above those usually found in the stainless steels) fall into the superalloy category and are discussed below.

Castings are classified as heat resistant if they are capable of sustained operation while exposed, either continuously or intermittently, to operating temperatures that result in metal temperatures in excess of 650 °C (1200 °F). Alloys used in castings for such service consist of five basic types - iron-chromium ("straight chromium"), iron-chromium-nickel, iron-nickel-chromium, nickel-base, and cobalt-base alloys.

For optimum materials selection of heat-resistant alloys, considerations include: (*a*) resistance to corrosion at elevated temperatures; (*b*) stability (resistance to warping, cracking, or thermal fatigue); and (*c*) creep strength (resistance to plastic flow). Many alloys of the same general type also are used for their resistance to corrosive mediums at temperatures below 650 °C (1200 °F). Castings intended for such service are classified as corrosion resistant. Although there is usually a distinction between heat-resistant alloys and corrosion-resistant alloys based on carbon content, the line of demarcation is vague, particularly for those alloys used at temperatures ranging from 480 to 650 °C (900 to 1200 °F).

Commercial applications of heat-resistant castings include metal treatment furnaces, gas turbines, aircraft engines, military equipment, oil refinery furnaces, cement mill equipment, petrochemical

Table 8.1 Compositions of ACI heat-resistant casting alloys

ACI designation	UNS number	ASTM specifications(a)	C	Composition, %(b) Cr	Ni	Si (max)
HA	...	A217	0.20 max	8 to 10	...	1.00
HC	J92605	A297, A608	0.50 max	26 to 30	4 max	2.00
HD	J93005	A297, A608	0.50 max	26 to 30	4 to 7	2.00
HE	J93403	A297, A608	0.20 to 0.50	26 to 30	8 to 11	2.00
HF	J92603	A297, A608	0.20 to 0.40	19 to 23	9 to 12	2.00
HH	J93503	A297, A608	0.20 to 0.50	24 to 28	11 to 14	2.00
HI	J94003	A297, A567, A608	0.20 to 0.50	26 to 30	14 to 18	2.00
HK	J94224	A297, A351, A567, A608	0.20 to 0.60	24 to 28	18 to 22	2.00
HL	J94604	A297, A608	0.20 to 0.60	28 to 32	18 to 22	2.00
HN	J94213	A297, A608	0.20 to 0.50	19 to 23	23 to 27	2.00
HP	...	A297	0.35 to 0.75	24 to 28	33 to 37	2.00
HP-50WZ (c)	...	...	0.45 to 0.55	24 to 28	33 to 37	2.50
HT	J94605	A297, A351, A567, A608	0.35 to 0.75	13 to 17	33 to 37	2.50
HU	...	A297, A608	0.35 to 0.75	17 to 21	37 to 41	2.50
HW	...	A297, A608	0.35 to 0.75	10 to 14	58 to 62	2.50
HX	...	A297, A608	0.35 to 0.75	15 to 19	64 to 68	2.50

(a) ASTM designations are same as ACI designations. (b) Rem Fe in all compositions. Manganese content: 0.35 to 0.65% for HA, 1% for HC, 1.5% for HD and 2% for the other alloys. Phosphorus and sulfur contents: 0.04% max for all but HP-50WZ. Molybdenum is intentionally added only to HA, which has 0.90 to 1.20% Mo; maximum for other alloys is set at 0.5% Mo. HH also contains 0.2% max N. (c) Also contains 4 to 6% W, 0.1 to 1.0% Zr, and 0.035% max S and P.

furnaces, chemical process equipment, power plant equipment, steel mill equipment, turbochargers, and equipment used in manufacturing glass and synthetic rubber. In general, alloys of the iron-chromium and iron-chromium-nickel groups are of the greatest commercial importance.

PHYSICAL AND MECHANICAL PROPERTIES

General characteristics of the five types of cast heat-resistant alloys are discussed below. Table 8.1 lists designations and compositions of iron-chromium, iron-chromium-nickel, and iron-nickel-chromium superalloys, whereas Table 8.2 gives typical room-temperature tensile properties. Table 8.3 presents corrosion rates in air and in flue gas. As previously indicated, some alloys discussed in Table 8.1, notably HA, HF, HH, and HK, are stainless steels rather than superalloys. Compositions of low-iron nickel-base and low-iron cobalt-base alloys are given in Tables 8.4 and 8.5, respectively. For more specific information on these alloys, see "Properties of Cast Heat-Resistant Alloys" in Volume 3 of the 9th Edition of *Metals Handbook*.

Iron-Chromium Alloys

Iron-chromium, or straight chromium, alloys contain 10 to 30% Cr and little or no nickel. These alloys are useful chiefly for resistance to oxidation. They have low strength at elevated temperatures, and their use is restricted to conditions, either oxidizing or reducing, that involve low static loads and uniform heating. Chromium content depends on anticipated service temperature. Again, most of the alloys in this group are actually stainless steels.

Table 8.2 Typical room-temperature properties of ACI heat-resistant castings alloys

Alloy	Condition	Tensile strength MPa	Tensile strength ksi	Yield strength MPa	Yield strength ksi	Elongation, %	Hardness, HB
HC	As cast	760	110	515	75	19	223
	Aged(a)	790	115	550	80	18	...
HD	As cast	585	85	330	48	16	90
HE	As cast	655	95	310	45	20	200
	Aged(a)	620	90	380	55	10	270
HF	As cast	635	92	310	45	38	165
	Aged(a)	690	100	345	50	25	190
HH, type 1	As cast	585	85	345	50	25	185
	Aged(a)	595	86	380	55	11	200
HH, type 2	As cast	550	80	275	40	15	180
	Aged(a)	635	92	310	45	8	200
HI	As cast	550	80	310	45	12	180
	Aged(a)	620	90	450	65	6	200
HK	As cast	515	75	345	50	17	170
	Aged(b)	585	85	345	50	10	190
HL	As cast	565	82	360	52	19	192
HN	As cast	470	68	260	38	13	160
HP	As cast	490	71	275	40	11	170
HT	As cast	485	70	275	40	10	180
	Aged(b)	515	75	310	45	5	200
HU	As cast	485	70	275	40	9	170
	Aged(c)	505	73	295	43	5	190
HW	As cast	470	68	250	36	4	185
	Aged(d)	580	84	360	52	4	205
HX	As cast	450	65	250	36	9	176
	Aged(c)	505	73	305	44	9	185

(a) Aging treatment: 24 h at 760 °C (1400 °F), furnace cool. (b) Aging treatment: 24 h at 760 °C (1400 °F), air cool. (c) Aging treatment: 48 h at 980 °C (1800 °F), air cool. (d) Aging treatment: 48 h at 980 °C (1800 °F), furnace cool.

Iron-Chromium-Nickel Alloys

Iron-chromium-nickel alloys contain more than 13% Cr and more than 7% Ni (always more chromium than nickel). These austenitic alloys ordinarily are used under oxidizing or reducing conditions similar to those withstood by the ferritic iron-chromium alloys. In service, however, they have greater strength and ductility than the straight chromium alloys. They are used, therefore, to withstand greater loads and moderate changes in temperature. These alloys also are used in the presence of oxidizing and reducing gases that are high in sulfur content. Of these alloys, the compositions that have 25% or less chromium and 20% or less nickel generally are considered stainless steels.

Iron-Nickel-Chromium Alloys

Iron-nickel-chromium alloys contain more than 25% Ni and more than 10% Cr (always more nickel than chromium). These austenitic alloys are used for withstanding reducing as well as oxidizing

Table 8.3 Approximate rates of corrosion for ACI heat-resistant casting alloys in air and in flue gas

Alloy	Oxidation rate in air, mils/yr(a) 870 °C (1600 °F)	980 °C (1800 °F)	1090 °C (2000 °F)	Corrosion rate, mils/yr(a)(b), in flue gas with sulfur content of: 0.12 g/m³ Oxidizing	0.12 g/m³ Reducing	2.3 g/m³ Oxidizing	2.3 g/m³ Reducing
HB.......	25 −	250 −	500 −	100 +	500 −	250 −	500
HC.......	10	50	50	25 −	25 +	25	25 −
HD.......	10 −	50 −	50 −	25 −	25 −	25 −	25 −
HE.......	5 −	25 −	35 −	25 −	25 −	25 −	25 −
HF.......	5 −	50 +	100	50 +	100 +	50 +	250 −
HH.......	5 −	25 −	50	25 −	25	25	25 −
HI	5 −	10 +	35 −	25 −	25 −	25 −	25 −
HK.......	10 −	10 −	35 −	25 −	25 −	25 −	25 −
HL.......	10 +	25 −	35	25 −	25 −	25 −	25 −
HN.......	5	10 +	50 −	25 −	25 −	25	25
HP.......	25 −	25	50	25 −	25 −	25 −	25 −
HT.......	5 −	10 +	50	25	25 −	25	100
HU.......	5 −	10 −	35 −	25 −	25 −	25 −	25
HW	5 −	10 −	35	25	25 −	50 −	250
HX.......	5 −	10 −	35 −	25 −	25 −	25 −	25 −

(a) Data based on 100-h tests. To convert to μm/yr, multiply by 25. (b) At 980 °C (1800 °F).

Sources: Alloy Casting Institute; A. Brasunas, J. T. Gow and O. E. Harder, "Resistance of Fe-Ni-Cr Alloys to Corrosion in Air at 1600 to 2200 F", ASTM Symposium on Materials for Gas Turbines, June 1946, p 129 to 152

atmospheres, except where sulfur content is appreciable. For atmospheres containing 0.05% or more hydrogen sulfide, for example, iron-chromium-nickel alloys are recommended.

In contrast with iron-chromium-nickel alloys, iron-nickel-chromium alloys do not carburize rapidly or become brittle and do not pick up nitrogen in nitriding atmospheres. These characteristics become enhanced as nickel content is increased, and in carburizing and nitriding atmospheres, casting life increases with nickel content.

Austenitic iron-nickel-chromium alloys are used extensively under conditions of severe temperature fluctuations, such as those encountered by fixtures used in quenching and by parts that are not heated uniformly or that are heated and cooled intermittently. Additionally, these alloys have characteristics that make them suitable for electrical resistance heating elements.

Nickel-Base Alloys

Nickel-base alloys contain about 50% Ni and appreciable amounts of chromium, cobalt, and refractory metals, but little or no iron. They may also contain aluminum and titanium. Originally, the high-chromium nickel-base alloys were developed for oxidation resistance, and those high in molybdenum were developed for chemical corrosion resistance. When aluminum and titanium are added to superalloys, high-temperature strength is increased significantly.

Nickel-base precipitation-strengthened superalloys exhibit elevated temperature mechanical properties that are superior to those of other

Table 8.4 Compositions of nickel-base superalloy castings

Alloy designation	Nominal composition, %										
	C	Ni	Cr	Co	Mo	Fe	Al	B	Ti	W	Zr	Others
B-1900	0.1	64	8	10	6	...	6	0.015	1	...	0.10	4Ta(a)
Hastelloy X	0.1	50	21	1	9	18	...	...	...	1	...	...
IN-100	0.18	60.5	10	15	3	...	5.5	0.01	5	...	0.06	1V
IN-738X	0.17	61.5	16	8.5	1.75	...	3.4	0.01	3.4	2.6	0.1	1.75Ta, 0.9Nb
IN-792	0.2	60	13	9	2.0	...	3.2	0.02	4.2	4	0.1	4Ta
Inconel 713C	0.12	74	12.5	...	4.2	...	6	0.012	0.8	...	0.1	2Nb
Inconel 713LC	0.05	75	12	...	4.5	...	6	0.01	0.6	...	0.1	2Nb
Inconel 718	0.04	53	19	...	3	18	0.5	...	0.9	...	...	0.1Cu, 5Nb
Inconel X-750	0.04	73	15	...	...	7	0.7	...	2.5	...	...	0.25Cu, 0.9Nb
M-252	0.15	56	20	10	10	...	1	0.005	2.6	...	...	...
MAR-M 200	0.15	59	9	10	...	1	5	0.015	2	12.5	0.05	1Nb(b)
MAR-M 246	0.15	60	9	10	2.5	...	5.5	0.015	1.5	10	0.05	1.5Ta
MAR-M 247	0.15	59	8.25	10	0.7	0.5	5.5	0.015	1	10	0.05	1.5Hf, 3Ta
NX 188 (DS)	0.04	74	...	...	18	...	8	...	...	...	...	...
René 77	0.07	58	15	15	4.2	...	4.3	0.015	3.3	...	0.04	...
René 80	0.17	60	14	9.5	4	...	3	0.015	5	4	0.03	...
René 100	0.18	61	9.5	15	3	...	5.5	0.015	4.2	...	0.06	1V
TRW-NASA VIA	0.13	61	6	7.5	2	...	5.5	0.02	1	6	0.13	0.4Hf, 0.5Nb, 0.5Re, 9Ta
Udimet 500	0.1	53	18	17	4	2	3	...	3	...	...	...
Udimet 700	0.1	53.5	15	18.5	5.25	...	4.25	0.03	3.5	...	...	...
Udimet 710	0.13	55	18	15	3	...	2.5	...	5	1.5	0.08	...
Waspaloy	0.07	57.5	19.5	13.5	4.2	1	1.2	0.005	3	...	0.09	...
WAZ-20 (DS)	0.20	72	...	...	...	...	6.5	...	...	20	1.5	...

(a) B-1900 + Hf also contains 1.5% Hf. (b) MAR-M 200 + Hf also contains 1.5% Hf.

superalloys. They are precision investment cast under vacuum in many configurations, but principally as turbine airfoils (blades and vanes). Nickel-base superalloys are more costly than iron-base alloys (based on the low cost of iron in comparison to nickel), but less expensive than cobalt-base alloys.

ASTM A560 describes two grades of chromium-nickel casting alloys: 50Cr-50Ni and 60Cr-40Ni. These two alloys (not strictly considered nickel-base alloys) are cast into tube supports and other firebox fittings for certain stationary and marine boilers. The greatest advantage of the chromium-nickel alloys is their resistance to hot slag corrosion in boilers that fire oil with high vanadium contents. Hot slag that is high in V_2O_5 content is extremely destructive to most other superalloys. For example, iron-base alloys frequently last less than one fourth as long as chromium-nickel alloys in regions of a firebox where the components are continually covered with molten slag containing V_2O_5.

Cobalt-Base Alloys

Cobalt-base alloys contain about 50% or more cobalt, plus appreciable amounts of chromium and refractory metals. These alloys have better high-temperature strength than the solution-strengthened iron-nickel-chromium alloys. Cobalt-base alloys are precision investment cast in many configurations. They are not as strong as nickel-base alloys in short-time tests, but are competitive in strength and corrosion resistance with nickel-base alloys at very high temperatures and for long periods of operation. At present, the major uses of these alloys are furnace fixtures and turbine vanes.

MANUFACTURE

Iron-base alloys can be cast from heats that are melted in electric arc furnaces that have either acid or basic linings. When melting is done in acid-lined furnaces, however, chromium losses are high, and silicon content is difficult to control. Thus, acid-lined furnaces are seldom used. All superalloys can be melted in high-frequency induction furnaces. Initial melting also can be done in consumable arc, electron beam, or other furnaces with appropriate atmosphere control. Superalloys that contain appreciable amounts of aluminum, titanium, or other reactive metals are melted by induction or electron beam processes under vacuum or a protective atmosphere prior to casting.

Iron-Base Alloy Castings

Iron-base alloy castings can be made by the static method, the centrifugal method, or the investment process. The essential feature of centrifugal casting that differentiates it from all the various static casting processes is the introduction of molten metal into a mold that is rotated during solidification. The investment process (sometimes called "lost wax") involves use of a wax pattern in the shape of the part that is invested (coated or surrounded) with a

Table 8.5 Compositions of cobalt-base superalloy castings

Alloy designation	Nominal composition, % C	Co	Cr	Ni	Al	B	Fe	Ta	W	Zr	Others
AiResist 13	0.45	62	21	...	3.4	...	...	2	11	...	0.1Y
AiResist 213	0.20	64	20	0.5	3.5	...	0.5	6.5	4.5	0.1	0.1Y
AiResist 215	0.35	63	19	0.5	4.3	...	0.5	7.5	4.5	0.1	0.1Y
Haynes 21	0.25	64	27	3	...	...	1	...	...	...	5 Mo
Haynes 25; L-605	0.1	54	20	10	...	...	1	...	15	...	...
Haynes 151(a)	0.48	65	20	...	...	0.03	...	...	12.8	...	3 max Fe + Ni
J-1650	0.20	36	19	27	...	0.02	...	2	12	...	3.8 Ti
MAR-M 302	0.85	58	21.5	...	...	0.005	0.5	9	10	0.2	...
MAR-M 322	1.0	60.5	21.5	...	...	...	0.5	4.5	9	2	0.75 Ti
MAR-M 509	0.6	54.5	23.5	10	...	...	...	3.5	7	0.5	0.2 Ti
MAR-M 918	0.05	52	20	20	...	...	...	7.5	...	0.1	...
NASA Co-W-Re	0.40	67.5	3	...	...	...	...	...	25	1	2 Re, 1 Ti
S-816	0.4	42	20	20	...	...	4	...	4	...	4 Mo, 4 Nb, 1.2 Mn, 0.4 Si
V-36	0.27	42	25	20	...	...	3	...	2	...	4 Mo, 2 Nb, 1 Mn, 0.4 Si
WI-52	0.45	63.5	21	...	...	...	2	...	11	...	2 Nb+Ta
X-40	0.50	57.5	22	10	...	...	1.5	...	7.5	...	0.5 Mn, 0.5 Si

(a) Obsolete alloy, included for reference purposes.

refractory. The wax pattern is then burned off (hence lost) to form the mold cavity into which the molten metal is poured. The centrifugal method is used extensively in production of tubular parts such as radiant tubes for furnaces.

With the exception of a small tonnage of iron-chromium alloys, iron-base superalloy castings usually are not heat treated before being shipped. Because the 12Cr and 18Cr alloys are hardenable, castings of these alloys sometimes require full annealing for removal of casting stresses.

Nickel-Base and Cobalt-Base Alloy Castings

Nickel-base and cobalt-base alloy castings generally are made by the investment casting process. Cobalt-base alloys usually are melted and cast in air, although some of the more advanced alloys such as MAR-M 509 must be melted and cast under vacuum. In the past, most nickel-base and cobalt-base castings were small, about 1 kg (2 lb) or less. However, with improved casting procedures and weld repair processes, in conjunction with the advent of hot isostatic pressing (HIP) as a tool to improve casting quality by sealing internal porosity, large complex investment cast parts - up to 150 cm (60 in.) in diameter - are replacing forgings and weldments in aircraft engines.

For many applications, both nickel-base and cobalt-base alloy castings must be diffusion coated with a material that is high in silicon or aluminum to enable the alloy to resist the service atmosphere. Cobalt-base alloys are used as cast, with or without a diffusion coating, but without any other thermal treatment. Nickel-base alloys may be used as cast; in the cast, coated, and aged condition; or in the cast, solution treated, coated, and aged condition. The last condition applies more generally to advanced nickel-base alloys.

Turbine vanes and blades cast in nickel-base alloys usually are shipped in the as-cast or solution treated condition. Turbine manufacturers perform coating and aging treatments on machined castings as part of their manufacturing operation.

MICROSTRUCTURES OF IRON-BASE ALLOYS

The microstructures of iron-chromium-nickel and iron-nickel-chromium alloys must be wholly austenitic or mostly austenitic with some ferrite, if these alloys are to be used for heat-resistant service. Depending on chromium and nickel contents, the microstructures of these iron-base alloys can be austenitic (stable), ferritic (stable, but also soft, weak, and ductile), or martensitic (unstable). Therefore, chromium and nickel contents should be selected to provide good strength at elevated temperatures, combined with resistance to carburization and hot gas corrosion. Ferritic and martensitic microstructures usually are associated with stainless steel compositions. The more highly alloyed superalloys generally exhibit the austenitic microstructure. However, some borderline compositions between the stainless steels and superalloys may exhibit microstructures of ferrite, or ferrite and austenite.

A fine dispersion of carbides or intermetallic compounds in an austenitic matrix increases high-temperature strength. For this reason, iron-base superalloys are higher in carbon content than corrosion-resistant steels. By holding at temperatures where carbon diffusion is rapid - such as above 1200 °C (2200 °F) - and then rapidly cooling, a high and uniform carbon content is established, and up to about 0.20% C is retained in the austenite. Some chromium carbides are present in the microstructures of alloys with carbon contents greater than 0.20%, regardless of solution treatment.

Castings develop considerable segregation as they freeze. In standard grades, either in the as-cast condition or after rapid cooling from a temperature near the melting point, much of the carbon is in supersaturated solid solution. Subsequent reheating precipitates excess carbides. The lower the reheating temperature, the slower the reaction and the finer the precipitated carbides. Fine carbides increase creep strength and decrease ductility. Intermetallic compounds such as Ni_3Al, if present, have a similar effect.

Reheating material containing precipitated carbides in the temperature range 980 to 1200 °C (1800 to 2200 °F) will agglomerate and spheroidize the carbides, which reduces creep strength and increases ductility. Above 1100 °C (2000 °F), so many of the fine carbides are dissolved or spheroidized that this strengthening mechanism loses its importance. For service above 1100 °C (2000 °F), certain proprietary alloys of the iron-nickel-chromium type have been developed. Alloys for this type of service contain tungsten to form tungsten carbides, which are more stable than chromium carbides at these temperatures.

Aging at a low temperature, such as 760 °C (1400 °F), where a fine, uniformly dispersed carbide precipitate will form, confers a high level of strength that is retained at temperatures up to those where agglomeration changes the character of the carbide precipitation (overaging temperatures). Solution heat treatment or quench annealing, followed by aging, is the treatment generally employed to attain maximum creep strength.

Ductility usually is reduced when strengthening occurs. However, in some alloys, the strengthening treatment corrects an unfavorable grain-boundary network of brittle carbides, and both properties benefit. Such treatment is costly and may warp castings excessively. Hence, this treatment is applied to superalloy castings only for the small percentage of applications in which the need for premium performance justifies the high cost.

Carbide networks at grain boundaries are generally undesirable in iron-base superalloys. Grain-boundary networks usually occur in very high carbon alloys, or in alloys that have cooled slowly through the high-temperature ranges where excess carbon in the austenite is rejected as grain-boundary networks rather than as dispersed particles. These networks confer brittleness in proportion to their continuity.

Carbide networks also provide paths for selective attack in some atmospheres and in certain molten salts. Therefore, it is advisable in some salt bath applications to sacrifice the high-temperature strength imparted by high carbon content and gain resistance to intergranular corrosion by specifying that carbon content be no greater than 0.08%.

Iron-Chromium Alloys

Iron-chromium alloys, also known as straight chromium alloys, contain either 9 or 28% Cr. HC and HD alloys are included among the straight chromium alloys, although they contain low levels of nickel.

HC alloy containing 28% Cr is a borderline composition between the stainless steels and the superalloys. (It probably should be classified as a high-chromium iron alloy.) HC alloy resists oxidation and the effects of high-sulfur flue gases at temperatures up to about 1100 °C (2000 °F). It is used for applications in which strength is not a consideration, or for applications in which only moderate loads are involved, at temperatures of about 650 °C (1200 °F). It also is used where appreciable nickel cannot be tolerated, as in very high sulfur atmospheres, or where nickel may act as an undesirable catalyst and destroy hydrocarbons by causing them to crack.

HC alloy, like the chromium corrosion-resistant steels, is ferritic at all temperatures. Its ductility and impact strength are very low at room temperature, and its creep strength is very low at elevated temperatures unless some nickel is present. In a variation of HC alloy that contains more than 2% Ni, substantial improvement in all three of these properties is obtained by increasing the nitrogen content to 0.15% or more.

HC alloy becomes embrittled when heated for prolonged periods at temperatures between 400 and 550 °C (750 and 1025 °F). It also exhibits low resistance to impact. The alloy is magnetic and has a low coefficient of thermal expansion, comparable to that of carbon steel. It has about eight times the electrical resistivity and about half the thermal conductivity of carbon steel. Its thermal conductivity, however, is roughly double the value for austenitic iron-chromium-nickel alloys.

HD alloy (28Cr-5Ni) is very similar in general properties to HC, except that its nickel content affords it somewhat greater strength at high temperatures. The high chromium content of this alloy makes it suitable for use in high-sulfur atmospheres.

HD alloy has a two-phase, ferrite-plus-austenite microstructure that is not hardenable by conventional heat treatment. Long exposure at 700 to 900 °C (1300 to 1650 °F), however, may result in considerable hardening and severe loss of room-temperature ductility through formation of the chromium-iron sigma (σ) phase. Ductility may be restored by heating uniformly to 980 °C (1800 °F) or higher, and then cooling rapidly to below 650 °C (1200 °F).

Iron-Chromium-Nickel Alloys

Ferrous alloys in which chromium content exceeds nickel content are made in compositions ranging from 20Cr-10Ni to 30Cr-20Ni. However, many of these compositions are either stainless steels, or are borderline alloys between stainless steels and superalloys.

HE alloy (28Cr-10Ni), a borderline alloy, has excellent resistance to corrosion at elevated temperatures. Because of its higher chromium content, it can be used at higher temperatures than HF alloy, which is a 20Cr-10Ni stainless steel, and is suitable for applications up to 1100 °C (2000 °F). This alloy is stronger and more ductile at room temperature than the straight-chromium alloys.

In the as-cast condition, HE alloy has a two-phase, austenite-plus-ferrite structure containing carbides. HE castings cannot be hardened by heat treatment. However, as with HD castings, long exposure to temperatures near 815 °C (1500 °F) will promote formation of σ phase and consequent embrittlement of the alloy at room temperature. The ductility of this alloy can be improved somewhat by quenching from about 1100 °C (2000 °F).

Castings of HE alloy have good machining and welding properties. Thermal expansion is about 50% greater than that of either carbon steel or the iron-chromium alloy HC. Thermal conductivity is much lower than for HD or HC, but electrical resistivity is about the same. HE alloy is weakly magnetic.

HI alloy (28Cr-15Ni) is similar to the HH stainless steel composition (26Cr-12Ni), but contains more nickel and chromium, thus making it a borderline alloy between stainless steels and superalloys. The higher chromium content makes HI more resistant to oxidation than HH, and the additional nickel serves to maintain good strength at high temperatures. Exhibiting adequate strength, ductility, and corrosion resistance, this alloy has been used extensively for retorts operating with an internal vacuum at a continuous temperature of 1175 °C (2150 °F). It has an essentially austenitic structure that contains carbides and that, depending on the exact composition, may or may not contain small amounts of ferrite. Service at temperatures ranging from 760 to 870 °C (1400 to 1500 °F) results in precipitation of finely dispersed carbide, which increases strength and decreases ductility at room temperature. At service temperatures greater than 1100 °C (2000 °F), however, carbides remain in solution, and room-temperature ductility is not impaired.

HL alloy (30Cr-20Ni) is similar to the HK stainless steel (26Cr-20Ni), but its higher chromium content makes it a borderline composition and gives it greater resistance to corrosion by hot gases, particularly those containing appreciable amounts of sulfur. Because essentially equivalent high-temperature strength can be obtained with either HK or HL, the superior corrosion resistance of HL makes it especially useful for service in which excessive scaling must be avoided. The as-cast and aged microstructures of HL alloy, as well as its physical properties and fabricating characteristics, are similar to those of HK.

Iron-Nickel-Chromium Alloys

Iron-nickel-chromium alloys generally have structures that are more stable than those of iron-base alloys in which chromium is the predominant alloying element. There is no evidence of an embrittling phase change in iron-nickel-chromium alloys that would impair their ability to withstand prolonged service at elevated temperatures. Experimental data indicate that composition limits are not critical; therefore, production of castings from these alloys does not require the close composition control necessary for making castings from iron-chromium-nickel alloys.

The following general observations should be considered in selection of iron-nickel-chromium alloys: (*a*) as nickel content is increased, the ability of the alloy to absorb carbon from a carburizing atmosphere decreases; (*b*) as nickel content is increased, tensile strength at elevated temperatures decreases somewhat, but resistance to thermal shock and thermal fatigue increases; (*c*) as chromium content is increased, resistance to oxidation and to corrosion in chemical environments increases; (*d*) as carbon content is increased, tensile strength at elevated temperatures increases; and (*e*) as silicon content is increased, tensile strength at elevated temperatures decreases, but resistance to carburization increases somewhat.

HN alloy (25Ni-20Cr) contains sufficient chromium for good high-temperature corrosion resistance. HN has mechanical properties somewhat similar to those of the much more widely used HT alloy, but has better ductility. It is used for highly stressed components at temperatures ranging from 980 to 1100 °C (1800 to 2000 °F). In several specialized applications (notably, brazing fixtures), it has given satisfactory service at temperatures from 1100 to 1150 °C (2000 to 2100 °F). HN alloy is austenitic at all temperatures. Its composition limits lie well within the stable austenite field. In the as-cast condition, it contains carbide areas, and additional fine carbides precipitate on aging. HN is not susceptible to σ phase formation, and increases in its carbon content are not especially detrimental to ductility.

HP, HT, HU, HW, and HX alloys together constitute about one third of the total production of heat-resistant alloy castings. When used for fixtures and trays for heat treating furnaces, which are subjected to rapid heating and cooling, these five high-nickel alloys have exhibited excellent service life. Because these compositions are not as readily carburized as iron-chromium-nickel alloys, they are used extensively for parts of carburizing furnaces. Because they form an adherent scale that does not flake off, castings of these alloys are also used in enameling applications where loose scale would be detrimental. In many respects, there are no sharp lines of demarcation among HP, HT, HU, HW, and HX alloys as far as service applications are concerned.

HP alloy (35Ni-26Cr) is related to HN and HT alloys, but is higher in alloy content. It contains the same amount of chromium, but more nickel than HK, and the same amount of nickel, but more chromium than HT. This combination of elements makes HP resistant to both oxidizing and carburizing atmospheres at high temperatures. HP alloy exhibits creep-rupture properties from 980 to 1100 °C (1800 to 2000 °F) that are comparable to, or better than, those of HK-40 and HN alloys.

HP alloy is austenitic at all temperatures and is not susceptible to σ phase. Its microstructure consists of massive primary carbides in an austenitic matrix. Additionally, fine secondary carbides are precipitated within the austenite grains upon exposure to elevated temperatures.

HT alloy (35Ni-17Cr) contains nearly equal amounts of iron and alloying elements. Its high nickel content enables it to resist the thermal shock of rapid heating and cooling. In addition, HT is resistant to high-temperature oxidation and carburization and has good strength at the temperatures ordinarily used for heat treating steel. Except in high-sulfur gases, and provided that limiting creep-stress values are not exceeded, it performs satisfactorily in oxidizing atmospheres at temperatures up to 1150 °C (2100 °F) and in reducing atmospheres at temperatures up to 1100 °C (2000 °F).

HT alloy is used widely for highly stressed parts in general heat-resistant applications. It has an austenitic structure containing carbides in amounts that vary with carbon content and thermal treatment. In the as-cast condition, it has large carbide areas at interdendritic boundaries. However, fine carbides precipitate within the grains after exposure to service temperature, causing a decrease in room-temperature ductility. Increases in carbon content may decrease the high-temperature ductility of the alloy. Silicon contents above about 1.6% provide additional protection against carburization, but at some sacrifice in elevated temperature strength. Its resistance to thermal shock can be increased by addition of up to 2% Nb.

HU alloy (39Ni-18Cr) is similar to HT, but its higher chromium and nickel contents give it greater resistance to corrosion by either oxidizing or reducing hot gases, including those that contain sulfur in amounts up to 2.3 g/m^3 (Table 8.3). Its high-temperature strength and resistance to carburization are essentially the same as those of HT, and thus its superior corrosion resistance makes it especially well suited for severe service involving high stress and/or rapid thermal cycling in combination with an aggressive environment.

HW alloy (60Ni-12Cr) is especially well suited for applications in which wide and/or rapid fluctuations in temperature are encountered. Additionally, HW exhibits excellent resistance to carburization and high-temperature oxidation. HW alloy has good strength at steel treating temperatures, although it is not as strong as HT. It performs satisfactorily at temperatures up to about 1120 °C (2050 °F) in strongly oxidizing atmospheres and up to 1040 °C (1900 °F) in oxidizing or reducing products of combustion, provided that sulfur is not present in the gas. The generally adherent nature of its oxide scale makes HW suitable for enameling furnace service, where even small flakes of dislodged scale could ruin the work in process.

HW alloy is used widely for intricate heat treating fixtures that are quenched with the load and for many other applications (such as furnace retorts and muffles) that involve thermal shock, steep temperature gradients, and high stresses. Its microstructure is austenitic and contains carbides in amounts that vary with carbon content and thermal treatment. In the as-cast condition, the microstructure consists of a continuous interdendritic network of elongated eutectic carbides. Upon prolonged exposure at service temperatures, the austenitic matrix becomes uniformly peppered with small carbide particles, except in the immediate vicinity of eutectic carbides. This change in microstructure is accompanied by an increase in room-temperature strength, but no change in ductility.

HX alloy (66Ni-17Cr) is similar to HW, but contains more nickel and chromium. Its higher chromium content gives it substantially better resistance to corrosion by hot gases (even sulfur-bearing gases), which permits it to be used in severe service applications at temperatures up to 1150 °C (2100 °F). However, it has been reported that HX alloy decarburizes rapidly at temperatures from 1100 to 1150 °C (2000 to 2100 °F). High-temperature strength, resistance to thermal fatigue, and resistance to carburization are essentially the same as for HW. Consequently, HX is suitable for the same general applications that demand good corrosion resistance. The as-cast and aged microstructures of HX, as well as its mechanical properties and fabricating characteristics, are similar to those of HW.

MICROSTRUCTURES OF NICKEL-BASE ALLOYS

Nickel-base superalloy casting alloys generally contain substantial levels of aluminum and titanium (Table 8.4). These elements strengthen the austenitic matrix through precipitation of the γ' phase $Ni_3(Al,Ti)$. Various ratios of aluminum and titanium are used in the different nickel-base superalloys. Generally, however, titanium atoms can replace aluminum atoms up to a ratio of three titanium to one aluminum without altering the crystallographic structure of γ'. When excess titanium is present, Ni_3Ti (η, eta phase) precipitates. Because γ' is coherent with the matrix, precipitation of this phase has a greater strengthening effect than precipitation of η phase.

In addition to the strengthening imparted by γ' precipitation, solid-solution strengthening is conferred by addition of refractory elements. Grain-boundary strengthening is achieved by additions of boron, zirconium, carbon, and hafnium. Hafnium also enhances grain-boundary ductility. Stress-rupture curves for various nickel-base superalloys are shown in Fig. 8.1.

The strength of these alloys is complemented by superior corrosion resistance, which is conferred by chromium and aluminum; titanium may be more favorable than aluminum under hot corrosion conditions. Coatings are used on most nickel alloys for temperatures exceeding about 815 °C (1500 °F) to provide adequate protection from oxidation and corrosion at these temperatures.

Nickel-base superalloy castings are produced by investment casting under vacuum, and improvements in properties have been made not only through control of composition, but also through more precise control of microstructure. A significant advance in microstructure control was the development of a columnar grain structure produced by directional solidification. Single crystals of nickel-base superalloys have been produced by directional solidification as well. The absence of grain boundaries permits elements such as carbon, zirconium, and boron to be deleted from the composition. The resulting increase in melting point in turn provides improved flexibility in alloy composition and heat treatment.

Extensive use of nickel alloy castings essentially began with Inconel 713. Alloys currently are available that can be used at temperatures up to about 1040 °C (1900 °F). Use of precision investment castings in aircraft turbine engines, particularly for turbine airfoils, has been the single most important driving force in the growth of the investment casting industry. Castings are intrinsically stronger than forgings at elevated temperatures. The coarse grain size of castings as compared to fine-grain forgings favors strength at very high temperatures. Additionally, casting compositions can be effectively tailored for high-temperature strength, inasmuch as forgeability characteristics are not applicable. As such, as aircraft engine operating temperatures increased, castings replaced forged airfoils to

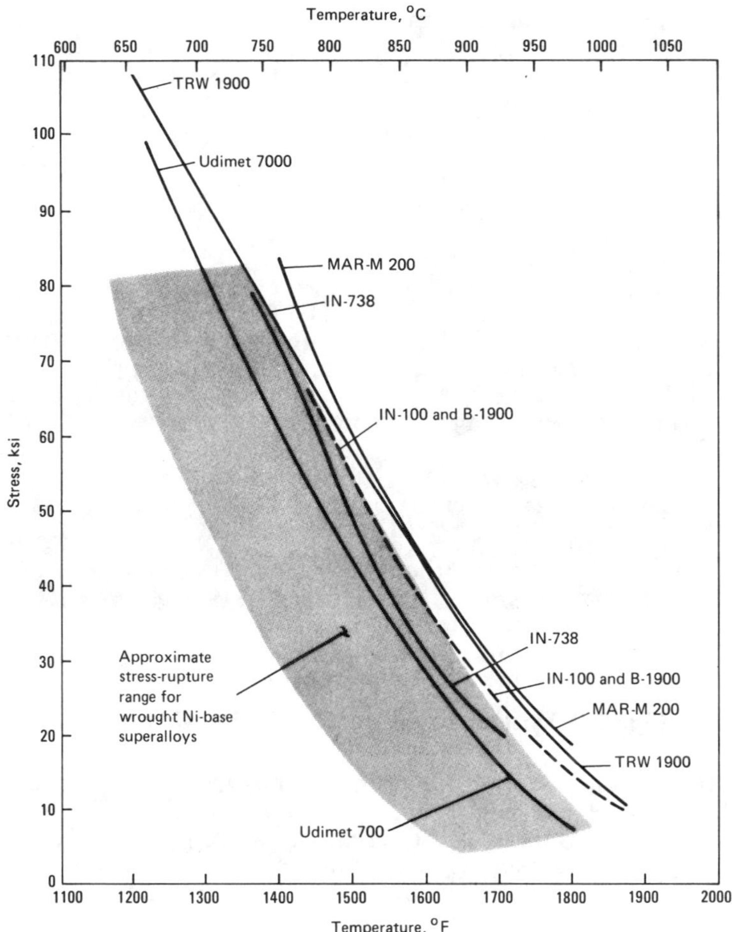

To convert stress values to MPa, multiply by 6.8948.

Fig. 8.1 Stress-rupture curves for 1000-h life of select nickel-base superalloys

provide satisfactory service durability. Further temperature increases have lead to use of air-cooled cast airfoil designs. A typical casting that is suitable for air cooling is illustrated in Fig. 8.2. For the air cooling process to be advantageous, complex casting designs are required.

In addition to creep strength and corrosion resistance, two other properties - stability and resistance to thermal fatigue - are important considerations in selection of nickel-base superalloy castings. Thermal fatigue resistance is controlled partially by composition, but it is also significantly affected by grain-boundary area and alignment relative to applied stresses. Crystallographic orientation of grains also influences thermal stresses, because the modulus of elasticity, which directly influences thermal stresses, varies with grain orientation. Stability of property values is influenced directly by metallurgical stability. Any microstructural

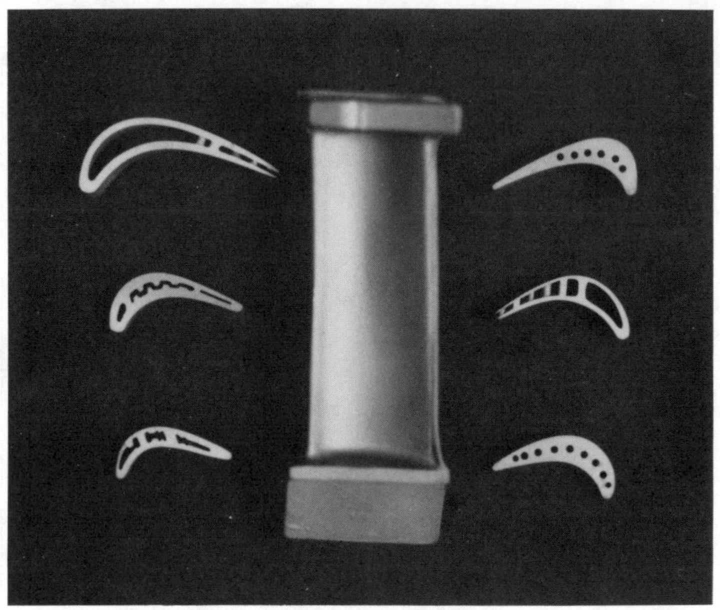

Note that the air cooling process requires complex castings to be effective.

Fig. 8.2 Typical casting suitable for air cooling

changes that occur during long-time exposure at high temperatures under stress cause attendant changes in properties. For instance, if the γ' phase coarsens, strength decreases. Also, potentially deleterious topologically close-packed (tcp) secondary phases, such as σ, Laves, and μ, may form. Coarsening of γ' can be controlled to some degree by adjusting alloy additions. Formation of tcp phases is controlled by adjusting the composition of the matrix to minimize the electron vacancy number (N_v). A high N_v indicates a tendency toward formation of tcp phases. In general, an N_v value below 2.4 indicates minimal formation of deleterious phases; however, this relationship varies with base-alloy composition.

Inconel 713C and 713LC are closely related investment casting alloys used principally for low-pressure turbine airfoils in gas turbines. Intended for operation at intermediate temperatures that range from 790 to 870 °C (1450 to 1600 °F), these alloys generally are used in uncooled airfoil designs.

IN-738X is an investment casting alloy with similar strength to Inconel 713C and Udimet 700, but with outstanding sulfidation resistance. It is used principally for latter stage turbine airfoils and for applications that are subject to hot corrosion, such as industrial and marine engines.

Udimet 700, although primarily a wrought alloy, also is used in investment cast high-pressure turbine blades. In cast form, it is similar in strength to Inconel 713C, but offers better hot corrosion resistance. It is designed for operation at intermediate temperatures from 730 to 900 °C (1350 to 1650 °F). A stability-controlled version of U-700 is known as Rene′ 77.

IN-100 is designed for use at metal temperatures up to about 980 °C (1800 °F) in cooled and uncooled airfoils. A stability-controlled version of IN-100 is known as Rene′ 100.

B-1900, to which 1% Hf usually is added to improve ductility and thermal fatigue resistance, is designed for use at metal temperatures up to about 980 °C (1800 °F) in cooled and uncooled airfoils.

Rene′ 80 offers excellent corrosion resistance in sulfur-bearing environments. It is designed for use at metal temperatures up to about 950 °C (1750 °F).

IN-792 is designed for use in applications similar to those of Rene′ 80. It is one of the most sulfidation-resistant nickel alloys available.

MAR-M 246 and MAR-M 247 are designed for use at metal temperatures of about 980 to 1010 °C (1800 to 1850 °F) in cooled and uncooled airfoils.

D.S. MAR-M 200 + Hf is produced by unidirectional solidification and is designed for metal temperatures of about 1010 to 1040 °C (1850 to 1900 °F). It is used in cooled airfoils.

Other alloys (such as Udimet 500) occasionally are used in turbine airfoil applications, and Inconel 718 has been cast into large static structures for gas turbines.

MICROSTRUCTURES OF COBALT-BASE ALLOYS

Cobalt-base superalloy castings were first used in highly stressed gas turbine blades during World War II. Although the initial alloy was used at temperatures no higher than those at which the iron-chromium-nickel and iron-nickel-chromium heat-resistant alloys were used, it gave satisfactory service under high stress and far surpassed the older alloys in creep-rupture properties. Compositions of select cobalt-base casting alloys are given in Table 8.5. Stress-rupture properties of commonly used cobalt-base alloys are shown in Fig. 8.3.

Most cobalt-base superalloys contain appreciable amounts of carbon and derive their strength not only from solid-solution strengthening by such elements as tungsten and chromium, but also from carbide precipitation, which becomes less effective at temperatures above about 815 °C (1500 °F). Nickel often is added to stabilize the high-temperature form of cobalt (face-centered cubic).

With the advent of vacuum melting, nickel-base alloys strengthened by precipitation of γ' phase promptly surpassed cobalt alloys in strength, and very few new cobalt-base casting alloys were developed for use in gas turbines after 1952. Generally, X-40 is used for latter stage turbine airfoils of older gas turbines, and MAR-M 509, WI-52, and MAR-M 302 are employed for first-stage, and occasionally second-stage, turbine vanes in gas turbines.

In terms of strength, cobalt alloys typically cannot compete with nickel alloys, except at temperatures above 980 °C (1800 °F). At these higher temperatures, the melting temperature advantage of cobalt over nickel is significant. However, because of the ease with which they can be repaired by welding and because of their high chromium contents, which give them good corrosion resistance at these high temperatures, cobalt alloys find extensive use in

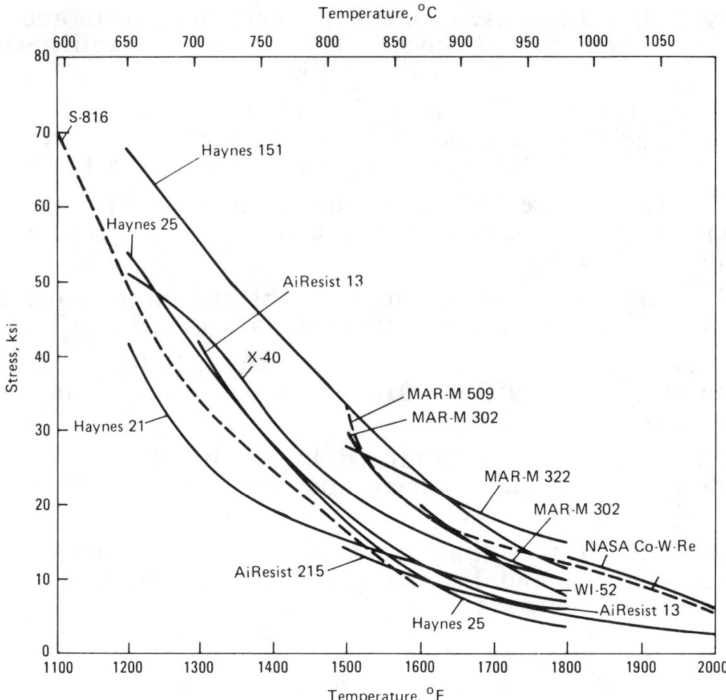

To convert stress values to MPa, multiply by 6.8948.

Fig. 8.3 Stress-rupture curves for 1000-h life of select cobalt-base alloys

high-pressure turbine vanes. Additional uses include cast components in burner cans and seals, and for some cobalt superalloys, castings for furnace hardware.

Five commonly used cobalt-base superalloys are described below. All are investment cast, usually from air-melted stock. None of the commercial alloys is directionally solidified. For proper materials selection of cobalt-base superalloys, stability and thermal fatigue resistance must be considered, because cobalt-base alloys generally are less resistant to thermal fatigue than nickel-base alloys of comparable strength. Stability can be inferred from the electron vacancy number (N_v). However, N_v numbers that delineate cutoff points for tcp phase formation tend to be near 2.6, which is higher than corresponding numbers for nickel-base alloys. All cobalt-base alloys exhibit aging on exposure due to the strengthening effects of changes in the carbide structure.

Haynes 21 and X-40 are used for turbine vanes in some gas turbines. They have good corrosion resistance and strength at temperatures from 700 to 815 °C (1300 to 1500 °F).

WI-52 is used for turbine vanes in some gas turbines. It can be used at temperatures up to about 950 °C (1750 °F), but must be coated for corrosion resistance.

Table 8.6 Machining data for ACI heat-resistant casting alloys

ACI designation	Typical hardness, HB	Rough turning(a) Speed, sfm(b)	Rough turning(a) Feed, ipr(c)	Finishing Speed, sfm(b)	Finishing Feed, ipr(c)	Drilling speed(d), sfm(b)
HA	220	40 to 50	0.010 to 0.030	80 to 100	0.005 to 0.010	35 to 70
HC	220	40 to 50	0.025 to 0.035	80 to 100	0.010 to 0.015	40 to 60
HD	190	40 to 50	0.025 to 0.035	80 to 100	0.010 to 0.015	40 to 60
HE	270	30 to 40	0.020 to 0.025	60 to 80	0.005 to 0.010	30 to 60
HF	190	25 to 35	0.020 to 0.025	50 to 70	0.005 to 0.010	20 to 40
HH	200	25 to 35	0.015 to 0.020	50 to 70	0.005 to 0.010	20 to 40
HI	200	25 to 35	0.015 to 0.020	50 to 70	0.005 to 0.010	20 to 40
HK	190	25 to 35	0.020 to 0.025	50 to 70	0.005 to 0.010	20 to 40
HL	190	30 to 40	0.020 to 0.025	60 to 80	0.005 to 0.010	30 to 60
HN	160	35 to 45	0.020 to 0.025	70 to 90	0.005 to 0.010	40 to 60
HP	...	35 to 45	0.020 to 0.025	70 to 90	0.005 to 0.010	40 to 60
HT	200	40 to 45	0.025 to 0.035	80 to 90	0.005 to 0.010	40 to 60
HU	190	40 to 45	0.025 to 0.035	80 to 90	0.010 to 0.015	40 to 60
HW	200	40 to 45	0.025 to 0.035	80 to 90	0.010 to 0.015	40 to 60
HX	185	40 to 45	0.025 to 0.035	80 to 90	0.010 to 0.015	40 to 60

(a) Single-point high speed steel tools usually are ground to 4 to 10° side and back rake, 4 to 7° side relief, 7 to 10° end relief, 8 to 15° end cutting-edge angle, 10 to 15° side cutting-edge angle, and 1/32- to 1/8-in. nose radius. (b) To convert to m/s, multiply by 0.005. (c) To convert to mm/rev, multiply by 25. (d) Recommended drilling feeds are as follows: for drill diameters up to 1/8 in., 0.001 to 0.002 ipr; 1/8 to 1/4 in., 0.002 to 0.004 ipr; 1/4 to 1/2 in., 0.004 to 0.007 ipr; 1/2 to 1 in., 0.007 to 0.015 ipr; over 1 in., 0.015 to 0.025 ipr. Tapping speeds recommended for HA, HC, HD, HE and HL are 10 to 25 sfm; for HF, HH, HI, and HK, 10 to 20 sfm; and for HN, HT, HU, HW and HX, 5 to 15 sfm.

MAR-M 302 is vacuum cast into turbine vanes for some gas turbines. This alloy is used at temperatures up to about 980 °C (1800 °F).

MAR-M 509 is vacuum cast into high-pressure turbine vanes and seals for gas turbines. It has good corrosion resistance and strength from 815 to 1010 °C (1500 to 1850 °F). MAR-M 509 generally is coated to provide the required corrosion resistance.

MATERIALS SELECTION

Superalloys are selected on the basis of structural integrity in a specific application. Strength, creep resistance, and corrosion resistance are prime factors influencing alloy selection. Next in importance is castability, although it is difficult to obtain a quantitative evaluation of this factor. The ability of an alloy to fill the mold must be assessed, as must the statistical distributions of product porosity and critical dimensions.

Because many castings must be machined to final dimensions, machinability of the as-cast metal must be evaluated. Quantitative machining data are available for many cast superalloys. Typical recommendations of speed, feed, and depth of cut are summarized in Table 8.6 for iron-base casting alloys. These recommendations must be considered starting points for evaluation of a particular machining operation. Adjustment of one or more of these factors will most likely be necessary to achieve optimum tool life. For further discussion of superalloy machining characteristics, see Chapter 12.

When total manufacturing cost and projected service life of a cast part have been determined, it is then possible to evaluate unit

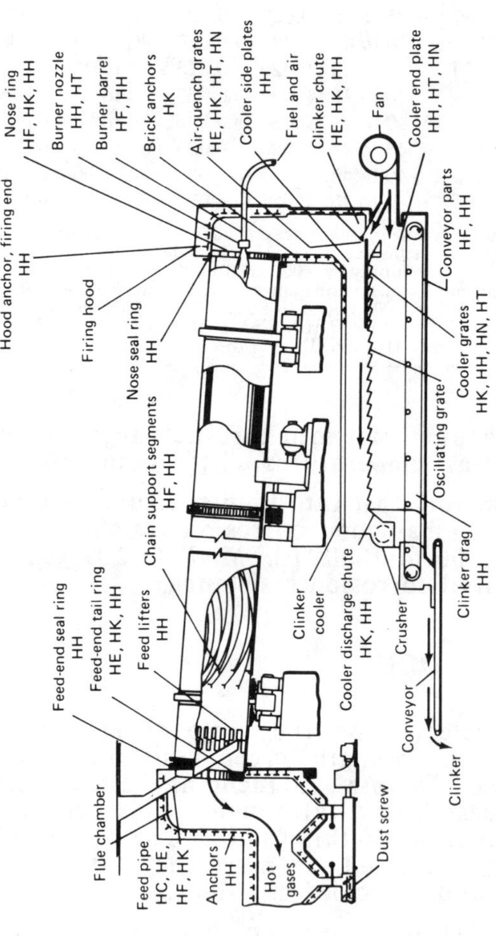

Hood anchor, firing end
HH
Firing hood
Nose seal ring
HH
Chain support segments
HF, HH
Nose ring
HF, HK, HH
Burner nozzle
HH, HT
Burner barrel
HF, HH
Brick anchors
HK
Air-quench grates
HE, HK, HT, HN
Cooler side plates
HH
Fuel and air
Clinker chute
HE, HK, HH
Fan
Cooler end plate
HH, HT, HN
Conveyor parts
HF, HH
Cooler grates
HK, HH, HN, HT
Oscillating grate
Clinker drag
HH
Crusher
Clinker
cooler
Cooler discharge chute
HK, HH
Conveyor
Clinker
Feed-end seal ring
HH
Feed-end tail ring
HE, HK, HH
Feed lifters
HH
Flue chamber
Feed pipe
HC, HE,
HF, HK
Anchors
HH
Hot
gases
Dust screw

Maximum operating temperature °C	°F	Part name	Environmental conditions	Alloys used(a)	Service life, yr
650	1200	Conveyor parts	Severe abrasion and oxidation	HF, HH_2	Indefinite
650	1200	Cooler discharge chute	Severe abrasion and oxidation	HH_2, HK	3 to 5
650	1200	Clinker drag	Severe abrasion and oxidation	HH_2	5 to 10
760	1400	Feed-end seal ring	Some abrasion and oxidation	HH_2	Indefinite
815	1500	Brick anchors	Even temperature	HK	Indefinite
815	1500	Burner barrel	Slight abrasion and oxidation	HF, HH_2	5 to 10
815	1500	Hood, anchor firing end	Even temperature, oxidation	HH_2	Indefinite
815	1500	Clinker chute	Severe abrasion, impact, oxidation	HE, HH_2, HK	Indefinite
815	1500	Air-quench grates	Severe abrasion and oxidation	HE, HK, HN, HT	3 to 7
925	1700	Anchors	Even temperature	HH_2	Indefinite
980	1800	Feed pipe	Moderate abrasion inside feed and dust particles outside, thermal shock, oxidation and sulfur gases	HC, HE, HF, HK	2 to 7
980	1800	Feed-end tail ring	Abrasive dust particles, thermal shock and oxidation	HE, HH_2, HK	10 to 15
980	1800	Feed lifters	Some abrasion, thermal shock, oxidation and sulfur gases	HH_2	5 to 10
980	1800	Chain support segments	Intermittent temperature surges, light abrasion, sulfur gases	HF, HH_2	Indefinite
980	1800	Cooler end plates	Severe abrasion and oxidation	HH_2, HN, HT	1 to 5
980	1800	Cooler grates	Severe abrasion and oxidation	HH_2, HK, HN, HT	1 to 5
980	1800	Cooler side plate	Severe abrasion and oxidation	HH_2	1 to 5
1100	2000	Nose seal ring	Some abrasion, oxidation and sulfur gases	HH_2	3 to 10
1100	2000	Burner nozzle	Some abrasion, oxidation and sulfur gases	HH_2, HT	1 to 3
1200	2200	Nose ring	Extreme abrasion, oxidation and sulfur gases	HF, HH_2, HK	3 to 5

(a) HH_2 is the type 2, wholly austenitic grade of iron-base 26Cr-12Ni alloy.

Fig. 8.4 Typical cement mill burning layout

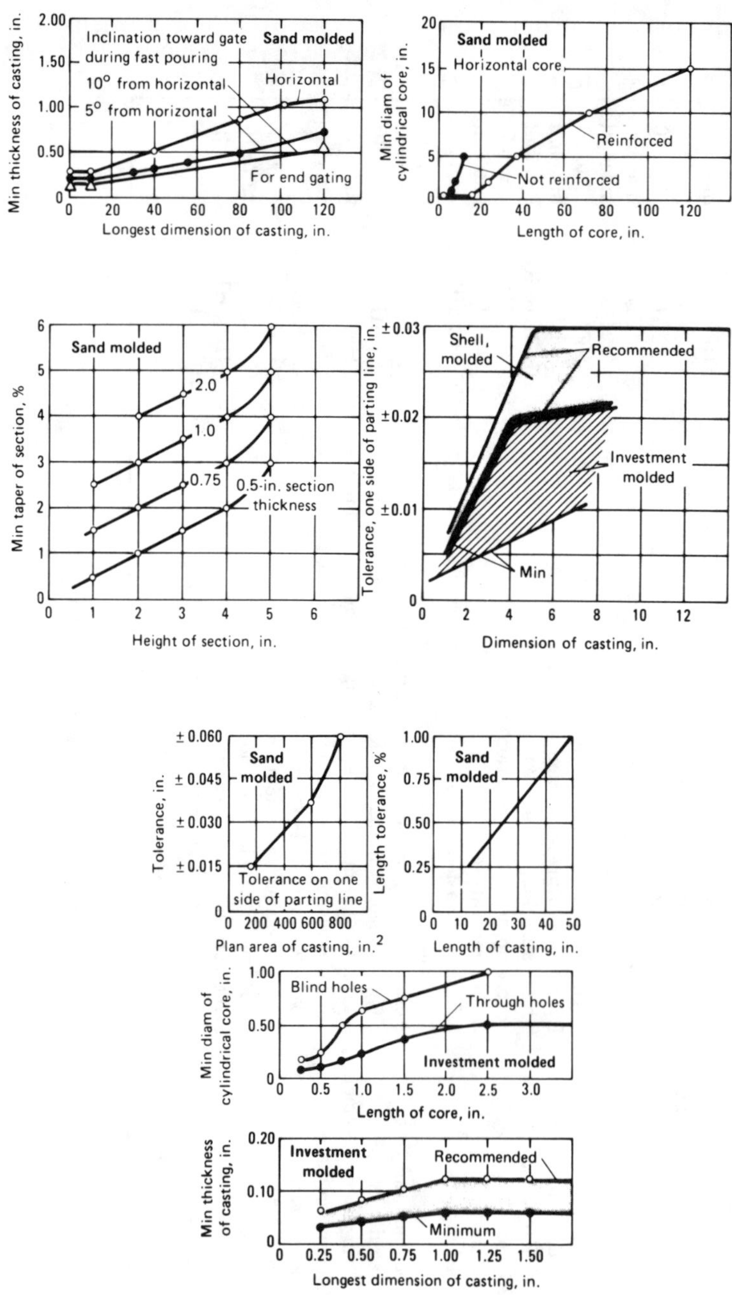

To convert dimensions in inches to equivalent values in millimeters, multiply by 25.

Fig. 8.5 Dimensional relationships in sand, shell, and investment molding for heat-resistant alloy castings

operating cost. Cost per hour of service life is the ultimate criterion in selection of a heat-resistant alloy. On this basis, an alloy that is more expensive in initial cost frequently provides lower cost per hour of service than a less-expensive type.

Typical cast heat-resistant alloys that are used for various components of a burning layout in a cement mill are shown in Fig. 8.4. More than one alloy can be used successfully in most of the twenty types of components listed. Environmental conditions that provide the basis for selection are listed for each component.

CASTING DESIGN

For most applications, ACI iron-base heat-resistant alloys are sand cast, although shell molding also is used. Sections in thicknesses of 4.8 mm ($^3/_{16}$ in.) and greater can be cast satisfactorily, and somewhat thinner sections also may be castable depending on casting design and pattern equipment. Dimensional tolerances for rough castings are influenced by the quality of the pattern equipment. In general, overall dimensions and locations of cored holes can be held to within 5.2 mm/m ($^1/_{16}$ in./ft).

The nickel-base and cobalt-base superalloys are usually investment cast. Dimensional tolerances for these castings are generally ±0.075 mm (±0.003 in.), with section thicknesses as low as 1.25 mm (0.05 in.) and excellent surface finish. More specific information on casting design is presented in Fig. 8.5.

Chapter 9

Forging

INTRODUCTION

Superalloys, because of their greater strength at elevated temperatures, are more difficult to forge than most other metals. Some of the iron-base superalloys, such as A-286, are similar to austenitic stainless steels in forgeability, but superalloys generally are more difficult to forge than stainless steels. Some superalloy compositions are so intrinsically strong at elevated forging temperatures that they cannot be shaped by conventional forging techniques. In these instances, the alloys are used either in the cast condition, or in wrought shapes made by powder metallurgy (P/M) processing.

FORGING METHODS

Forgeable superalloys are produced by open-die or closed-die forging, upsetting, extrusion forging, roll forging, or ring rolling. Often, two or more of these methods are used in sequence.

Open-Die Forging

Open-die forging (hand or flat-die forging) frequently is used to produce preforms for relatively large parts, such as disks and shafts for gas turbines. Many such preforms are completed in closed dies. Open-die forging is seldom used for producing forgings weighing less than 9 kg (20 lb).

Closed-Die Forging

Closed-die forging is used widely for forging superalloys. The procedures, however, are generally different from those used for forging similar shapes from carbon or low-alloy steels. For example, preforms made by open-die forging, upsetting, rolling, or extruding are used to a greater extent for closed-die forging of superalloys than for steel. Because of the greater difficulties in forging superalloys, compared with forging similar sizes and shapes from steel, diemaking is also different.

Upset Forging

Upset forging commonly is applied to superalloys, sometimes as the only forging operation, but more often to produce preforms, such as for turbine engine airfoils. In upset forging of superalloys, the maximum unsupported length of upset is about two diameters.

Extrusion Forging

Extrusion forging also is used to produce preforms for subsequent forging in closed dies and often competes with upsetting. Whether the preform is produced by extruding a slug or by forming an upset on the end of a smaller cross section depends primarily on the equipment available.

Roll Forging

Roll forging sometimes is used to produce preforms for subsequent forging in closed dies. The rolling techniques for preforming superalloys are basically the same as those for preforming steel. Roll forging saves material and decreases the number of closed-die operations required.

Ring Rolling

Ring rolling is sometimes used to save material when producing ringlike parts from hollow billets. The general method used for superalloys is essentially the same as that used for steel. Superalloys with forgeability ratings of 1 or 2 (Table 9.1) can be ring rolled with the same procedures as those used for carbon and low-alloy steels. Alloys with forgeability ratings of 3, 4, and 5 require more steps in ring rolling and supplemental heating with auxiliary torches.

FORGING RATINGS/TEMPERATURES

Forgeability ratings and forging temperatures of the most commonly forged superalloys are listed in Table 9.1. The forging temperatures given in Table 9.1 are the temperatures of the billets as they are removed from the furnace. Forging should begin immediately, with a loss in temperature of no more than 42 °C (75 °F). Forging can be continued until the stock has cooled 110 °C (200 °F), or more for some alloys, below the temperatures given in Table 9.1, without damage to the work metal. However, because greater pressures are required, forging is seldom done at temperatures substantially lower than those given in Table 9.1.

Table 9.1 Forging temperatures and forgeability ratings for superalloys

	Forging temperature(a)				
	Upset and breakdown		Finish forging		
Alloy	°C	°F	°C	°F	Forgeability rating(b)
Iron-base alloys					
A-286	1095	2000	1035	1900	1
V-57	1095	2000	1035	1900	1
16-25-6	1095	2000	1095	2000	1
Nickel-base alloys					
Alloy R-235	1205	2200	1205	2200	3
Astroloy	1120	2050	1120	2050	5
Hastelloy W	1205	2200	1035	1900	4
Hastelloy X	1175	2150	1175	2150	3
Inconel 600	1150	2100	1035	1900	1
Inconel 700	1120	2050	1105	2025	4
Inconel 718	1095	2000	1035	1900	2
Inconel X-750	1175	2150	1120	2050	2
Inconel 751	1150	2100	1150	2100	3
Incoloy 901	1150	2100	1095	2000	2
M-252	1150	2100	1095	2000	3
Rene' 41	1150	2100	1120	2050	3
U-500	1175	2150	1175	2150	4
U-700	1120	2050	1120	2050	5
Waspaloy	1160	2125	1035	1900	3
Cobalt-base alloys					
J-1570	1175	2150	1175	2150	2
J-1650	1150	2100	1150	2100	2
HS-25 (L-605)	1230	2250	1230	2250	3
S-816	1150	2100	1150	2100	4
Haynes 188	1205	2200	1205	2200	3

(a) Lower temperatures are often used for specific forgings when structural uniformity is a requirement. (b) Based on the considerations discussed in text. As the rating increases, forgeability decreases.

Forgeability Ratings

In the forgeability ratings listed in Table 9.1, alloy A-286 is assigned an arbitrary value of 1, because it is one of the most forgeable of the superalloys. The other alloys listed are assigned values that are multiples of 1, depending on their forgeability - as the arbitrary number increases, forgeability decreases. In establishing forgeability ratings, the power required is a minor consideration. Forgeability is determined by the ease with which a given shape can be formed. Alloys that are difficult to forge generally require more blows and, consequently, more operations than the more easily forgeable alloys. The values of 1 to 5 in Table 9.1 are based on the difficulties that are encountered in finishing the forging operation.

Rate of die deterioration, number of rejected forgings, and number of blows required to produce a given shape increase as forgeability decreases. These factors have been considered in establishing the ratings given in Table 9.1.

IRON-BASE ALLOYS

Stock for forgings of iron-base alloys generally is furnished as press-forged squares or hot rolled rounds, depending on size. As-cast ingots sometimes are used. Inclusion content of the alloys has a significant effect on forgeability. Alloys containing titanium and aluminum can develop nitride and carbonitride segregation, which later appears as stringers in wrought bars, especially in air-melted heats. Stringers impair forgeability. However, through improved vacuum melting techniques, this type of segregation has been almost completely eliminated. Thus, iron-base alloys can be forged into a greater variety of shapes with greater reductions, approaching the forgeability of type 304 stainless steel.

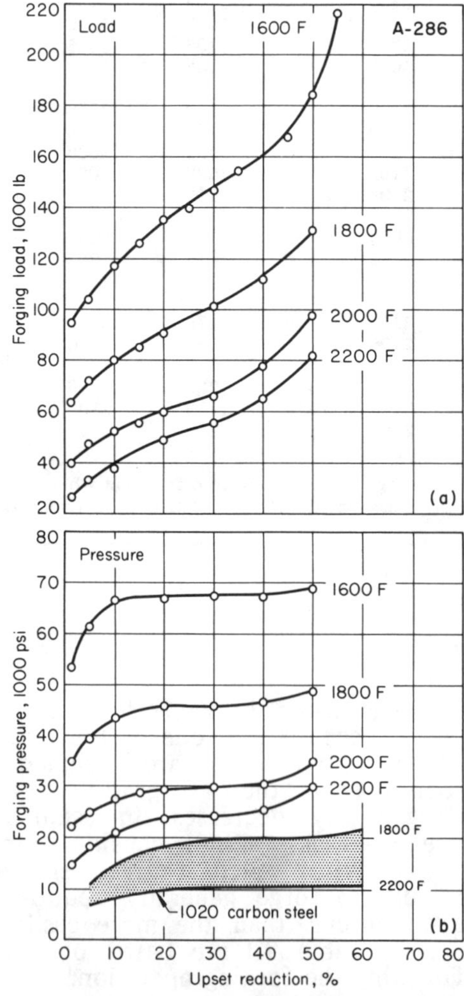

Fig. 9.1 Effect of upset reduction at four temperatures on (a) forging load in forging A-286 and (b) forging pressure for A-286, compared with that for 1020 carbon steel

Temperature has an important effect on forgeability. The optimum temperature range for forging A-286 and similar iron-base alloys is narrow. Forgeability of A-286, based on forging load required for various upset reductions at four forging temperatures, is shown in Fig. 9.1(a). Figure 9.1(b) shows that, on the basis of forging pressure, A-286 is considerably more difficult to forge than 1020 steel, even though A-286 is among the most forgeable of the heat-resistant alloys (Table 9.1). For instance, as shown in Fig. 9.1(b), 1020 steel, when heated to 1205 °C (2200 °F), requires only about 69 MPa (10 ksi) for an upset reduction of 30%, whereas for the same reduction, A-286 at 1205 °C (2200 °F) requires approximately 172 MPa (25 ksi).

Figure 9.2 shows that, on the basis of the specific energy required for various percentages of upset reduction, A-286 requires approximately 50% more energy than 4340 steel when both are forged at 1095 °C (2000 °F). However, A-286 requires only about half the specific energy that Rene′ 41 requires for the same upset reduction and the same forging temperature (Fig. 9.2).

Forging pressures increase considerably for greater upset reductions at normal forging temperatures. As shown in Fig. 9.3, the pressure for a 20% upset reduction of alloy A-286 at 1095 °C (2000 °F) is

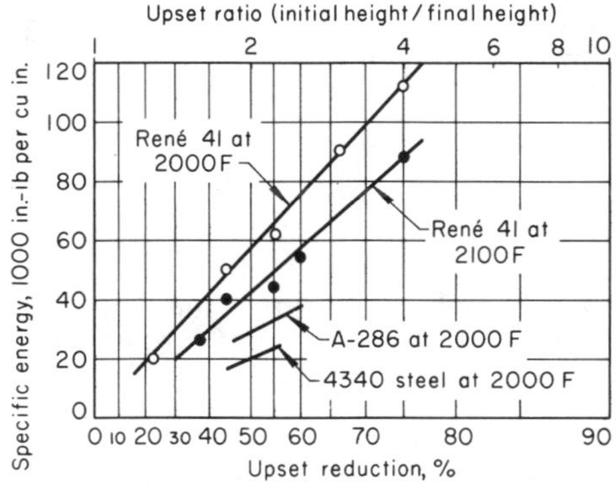

Fig. 9.2 Effect of upset reduction and forging temperature on specific energy required for forging two superalloys and 4340 steel

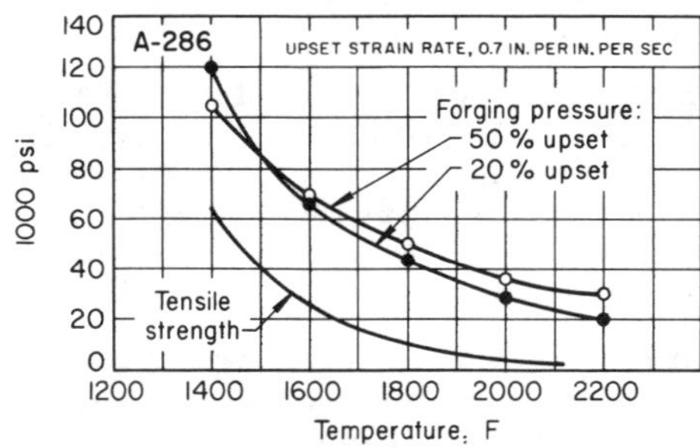

Fig. 9.3 Effect of temperature on forging pressure and tensile strength of A-286

about 193 MPa (28 ksi), but for an upset reduction of 50% the pressure increases to about 241 MPa (35 ksi). Figure 9.3 also shows that forging pressure is up to 10 or 12 times greater than the tensile strength of the alloy at forging temperature.

Strain rates also influence forging pressures. Figure 9.4 shows that as strain rate increases more energy is required in presses and hammers.

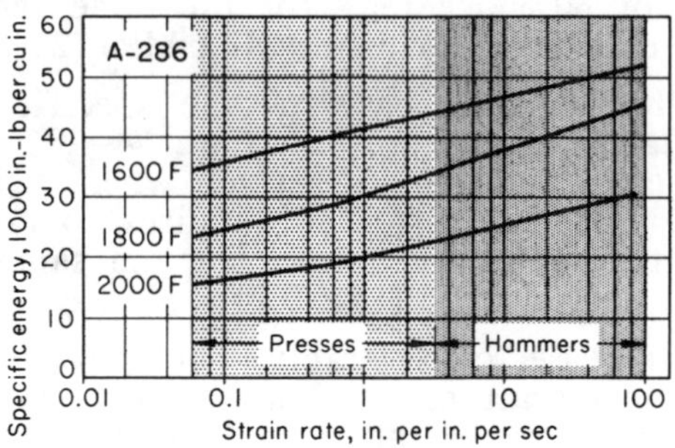

Fig. 9.4 Effect of strain rate on specific-energy requirements in press and hammer forging of A-286 at three different temperatures

NICKEL-BASE ALLOYS

With the development of vacuum melting technology, nickel superalloys have been produced that contain relatively larger amounts of the reactive elements, particularly titanium and aluminum for gamma prime (γ') precipitation strengthening. To improve the high-temperature strength of the forgeable nickel superalloys, titanium and aluminum contents have been increased, and chromium content has been reduced – the lower the chromium content, the higher the γ' solid-solution temperature. As high-temperature strength is increased, forgeability is decreased. To demonstrate this relationship, the following alloys are listed in order of increasing high-temperature strength and decreasing forgeability. The combined titanium and aluminum as well as chromium contents for these alloys are given in Table 9.2.

The Astroloy composition exhibits about the highest elevated temperature strength possible, while maintaining forgeability. Vacuum melting, of course, reduces the levels of oxygen and nitrogen, thereby eliminating most of the oxides and nitrides that contribute to poor forgeability. Because of this phenomenon, alloys with such good high-temperature strength as Astroloy are still forgeable.

Nickel-base alloys are available in various cogged billet and bar sizes for forging. These alloys commonly are melted by one of the following methods: (*a*) air melting followed by either vacuum induction melting or vacuum consumable electrode arc melting; (*b*) vacuum induction melting followed by vacuum consumable electrode arc melting; or (*c*) consumable electrode arc melting under slag.

Table 9.2 Relationship of forgeability and high-temperature strength of select nickel-base superalloys

Alloy	Ti + Al, %	Cr, %
Waspaloy	4.4	19.5
Rene'	4.5	19
U-500	6	18
Astroloy	7.5	15

Note: Alloys are listed in order of increasing high-temperature strength and decreasing forgeability.

Compared with ordinary arc-melting techniques, these three melting procedures have produced marked improvements in forgeability by reducing the levels of segregation. However, most ingots made on a production basis still contain enough segregation to influence forgeability. Because ingots produced by vacuum induction melting solidify progressively toward the center and take longer to freeze, the alloying elements and impurities concentrate at the center. Segregation in ingots produced by consumable electrode arc melting is generally less.

As shown in Table 9.1, all but one of the nickel alloys are less forgeable than the iron-base alloys - almost all require more force to produce a given shape. Astroloy and U-700 are the two most difficult-to-forge nickel-base alloys. For a given percentage of upset reduction at a forging temperature of 1095 °C (2000 °F), these alloys require about twice the specific energy needed for the iron-base alloy A-286. (See Fig. 9.2.)

In the forgeability ratings listed in Table 9.1, Astroloy and U-700 alloys have about one fifth the forgeability of Inconel 600. However, these ratings reflect only a relative ability to withstand deformation without failure; they do not indicate the energy or pressure needed for forging, nor can the ratings be related to low-alloy steels and other alloys that are considerably more forgeable.

Forging of nickel-base alloys requires close control over both metallurgical and operational conditions. Particular attention must be given to control of the work-metal temperature. Observers are usually required to record data on transfer time, soaking time, finishing temperature, and percentage reduction. Critical parts frequently are numbered, and precise records are kept. These records are useful in determining the cause of defective forgings, and they permit metallurgical analysis so that defects can be avoided in future products.

Nickel-base alloys are sensitive to minor variations in composition, which can cause large variations in forgeability, grain size, and final properties. In one instance, wide heat-to-heat variations in grain size occurred in parts forged from Incoloy 901 in the same sets of dies. For some parts, optimum forging temperatures had to be determined for each incoming heat of material, by making sample forgings and examining them after heat treatment for variations in grain size and other properties.

In the forging of nickel-base alloys, the forging techniques developed for one shape usually must be modified when another shape is forged from the same alloy; therefore, development time is often needed to establish suitable forging and heat treating cycles. This is especially true for alloys such as Waspaloy, Rene′ 41, U-500, and U-700.

COBALT-BASE ALLOYS

Many of the cobalt-base alloys cannot be forged successfully, because they ordinarily contain more carbon than iron-base alloys and, therefore, greater quantities of hard carbides, which impair forgeability. The four cobalt-base alloys listed in Table 9.1 are forgeable. The strength of these alloys at elevated temperatures, including the temperatures at which they are forged, is considerably higher than for iron-base alloys. Consequently, the pressures required to forge them are several times greater than those required for iron-base alloys.

Even when forged at their maximum forging temperature, alloys S-816 and HS-25 work harden; thus, forging pressure must be increased with greater reductions. Accordingly, these alloys generally require frequent reheating during forging to promote recrystallization and to lower the forging pressure for succeeding steps.

Forging conditions (temperature and reduction) have a significant effect on the grain size of cobalt-base alloys. Because low ductility, notch brittleness, and low fatigue strength are associated with coarse grains, close control of forging and of final heat treatment is important.

Cobalt-base alloys are susceptible to grain growth when heated above about 1175 °C (2150 °F). They heat slowly and require a long soaking time for temperature uniformity. Forging temperatures and reductions, therefore, depend on the forging operation and the part design.

These alloys usually are forged with small reductions in initial breakdown operations. The reductions are selected to impart sufficient strain to the metal so that recrystallization (and usually grain refinement) will occur during subsequent reheating. Because the cross section of a partly forged section has been reduced, less time is required to reach temperature uniformity in reheating. Consequently, because reheating time is shorter, the reheating temperature may sometimes be increased 28 to 84 °C (50 to 150 °F) above the initial forging temperature without damaging effects. However, if the part receives only small reductions in subsequent forging steps, forging should be continued at the lower temperatures. These small reductions, in turn, must be in excess of about 5 to 15% to prevent abnormal grain growth during subsequent annealing. The forging temperatures given in Table 9.1 are usually satisfactory.

RECENT DEVELOPMENTS

Superalloy forging development programs have been directed at two targets: improved mechanical properties and reduced costs. Improved mechanical properties are achieved by thermal mechanical working during the forging process and retention of strain strengthening by careful control of subsequent heat treatments. Most cost reductions have been related to more efficient utilization of material via near-net shape processing.

One significant result to date is the development of the isothermal forging process (Pratt & Whitney Aircraft gatorizing process). In this operation, the alloy to be forged is placed in a temporary condition of low strength and high ductility (superplastic) at the forging temperature. Forging occurs isothermally as both the dies and forging stock are heated to the established forging temperature and maintained at that temperature during forging. The process presently utilizes extrusion to consolidate powder into a fine-grain bar that is cut into billet size for forging. Powder metallurgy techniques are discussed in detail in Chapter 10.

The extrusion process introduces sufficient stress in the billet for superplastic conditions at the forging temperature. By use of superplastic isothermal forging techniques and P/M processing, nickel superalloy compositions that usually are used only in the cast condition can be forged.

The first application of the isothermal forging process to volume production of parts has been for the high thrust-to-weight F-100 military turbine engine. By utilizing an all inert P/M approach in conjunction with the isothermal forging process, compressor and turbine disks and spacers were produced from the high-strength IN-100 alloy (normally only used as castings) for the F-100 engine with a high level of uniformity of mechanical properties and a virtual absence of segregation.

For more detailed information on forging machines, dies, preparation of stock, heating of dies, use of lubricants, and cleaning, see "Forging of Heat-Resisting Alloys" in Volume 5 of the 8th Edition of *Metals Handbook*.

Chapter 10

Powder Metallurgy Processing

INTRODUCTION

The fabrication of fully dense nickel-base superalloy shapes by powder metallurgy (P/M) processing has become increasingly attractive. Interest in superalloy P/M processing is due to several reasons. First, advanced superalloys are prone to severe macrosegregation, which inhibits successful ingot breakdown. Conceptually, powder metallurgy offers a method for overcoming this problem. Because the material is divided into small droplets while it is a homogeneous liquid, the maximum segregation distance is reduced to the size of the solidified droplet. Furthermore, there is the possibility that P/M processing may offer improved economics through near-net shapes and fewer processing steps than required for ingot technology. Finally, P/M processing provides an opportunity to produce alloys with fully dense wrought structures that cannot be forged by conventional means.

Early efforts to process superalloys were frustrated by the oxidation of powders during processing, which led to poor tensile and rupture ductility. A breakthrough came with the inception of inert powder processing. In this process, powder production, collection, and densification are carried out in an inert atmosphere. The necessity of producing prealloyed powders of ever increasing purity brought about the development of atomization processes that operate in an inert atmosphere or in a vacuum.

Government-supported research in superalloy P/M technology has led to the P/M processing of many critical jet engine components. Superalloy parts made by P/M techniques are being used in advanced turbine engines because of the following advantages:

* Ability to produce near-net shapes, which results in reduced material input and less machining to produce a finished part
* Improved property uniformity and alloy development flexibility due to the elimination of macrosegregation
* Reduced energy requirements and shorter delivery time because the P/M process requires fewer processing steps than conventional ingot technology
* Ability to produce high-strength superalloys such as IN-100, which are usually only available as castings in fully dense wrought structures

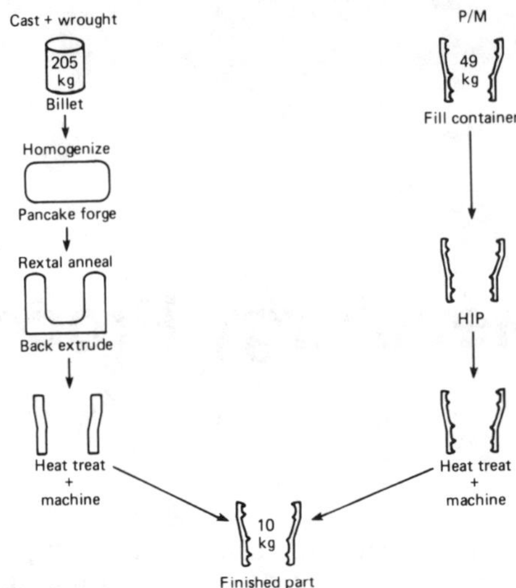

Fig. 10.1 Schematic comparison of processes required for manufacture of turbine engine shaft by cast-plus-wrought and P/M procedures

The superalloy engine shaft shown schematically in Fig. 10.1 is an example of the material and fabrication savings attainable with P/M technology. It is noteworthy that the P/M process utilizes markedly fewer processing steps and reduces the material input weight from 205 kg (452 lb) to 49 kg (108 lb).

For additional sources of information on P/M processing, see "Superalloy Powder Metallurgy," *Metals Handbook*; *Desk Edition.*

P/M SUPERALLOY PROCESS

The process for producing P/M superalloy hardware generally involves production of spherical prealloyed powder, screening to remove oversize particles, blending the powders to homogenize the powder size distribution, loading the powder into containers, vacuum outgassing and sealing the containers, and then consolidating the powder to full density.

The principal commercial powdermaking processes used for superalloys are:

* Inert-gas atomization
* Vacuum atomization
* Plasma rotating-electrode process (PREP)

In the gas-atomization process used for superalloys, metal is vacuum induction melted and subsequently atomized using an inert gas, usually argon. In the vacuum-atomization process for superalloys, molten metal is saturated with hydrogen. The metal is then streamed

Table 10.1 Prealloyed powder production methods for superalloys

	Commercial process			Laboratory scale and R&D processes		
Process	Argon atomization	Vacuum atomization	Rotating electrode	Centrifugal shot casting	Electron beam rotating ingot (PSV)	Electron beam rotating disk
Atomization technique	Disintegration of molten stream by argon jet	Exposure of hydrogen-containing melt to vacuum	Centrifugal atomization of electrode	Atomization from rotating crucible	Atomization from rotating ingot	Atomization from rotating disk
Melt source	Induction melting in ceramic crucible	Induction melting in ceramic crucible	Nonconsumable arc or transferred plasma arc	Consumable arc	Electron beam	Electron beam
Environment	Argon	Vacuum, hydrogen, and argon	Vacuum, argon, or helium	Vacuum, argon, or helium	Vacuum	Vacuum
Mean size, μm	..	10-50	225	300-400	400	...
Approximate yield, %	70	...	95	60	85	60
Comments	150-μm particles, cooling rate of ~10^{20} °C/s	...	Plasma arc avoids tungsten contamination	Cooling rate of 300-μm particle in helium, ~5×10^3 °C/s	...	...

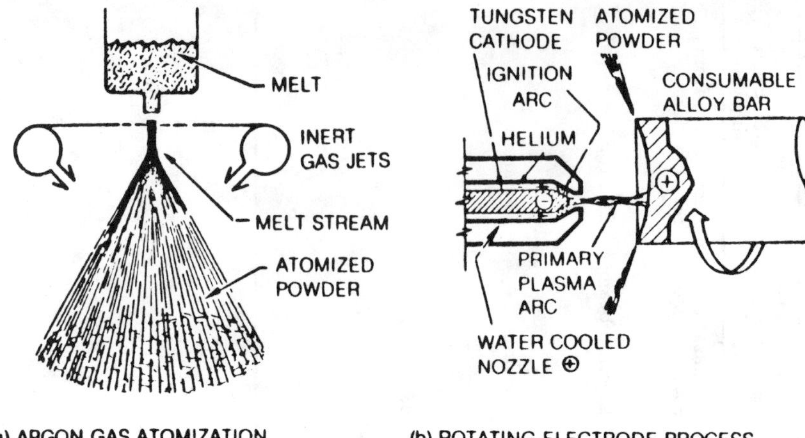

(a) ARGON GAS ATOMIZATION

(b) ROTATING ELECTRODE PROCESS

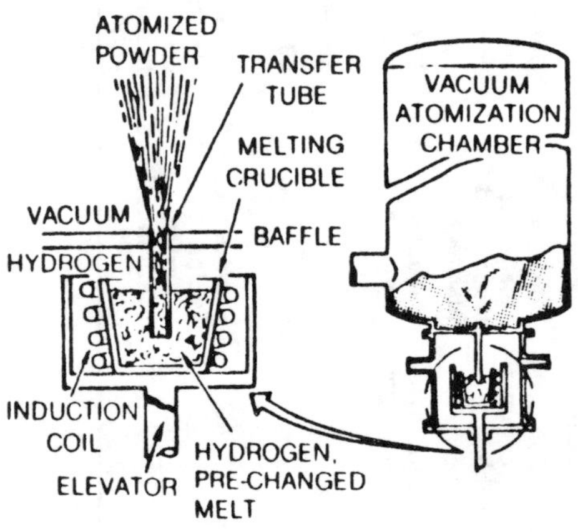

(c) SOLUBLE GAS PROCESS

Fig. 10.2 Commercial powder production processes

into a vacuum. As the gas expands, it comes out of solution and atomizes the metal. In the plasma rotating-electrode process, a plasma arc provides localized melting at the end of a rapidly rotating bar. The rotating motion spins off the powder from the molten pool.

Another atomization process that is still in the pilot stage is centrifugal atomization utilizing a rapidly spinning crucible, disk, or plate. In this process, molten alloy streams are poured onto a spinning disk, crucible, or plate. Droplets are ejected from the spinning base. Other pilot processes involve atomization from rotating ingots or disks using electron beams as the melt source. See Table 10.1 and Fig. 10.2 for typical prealloyed powder production methods.

Most superalloy powder produced currently is spherical and has a tap density of about 65%. Tap density is important for reliable production of P/M shapes. Figure 10.3 shows powder produced by

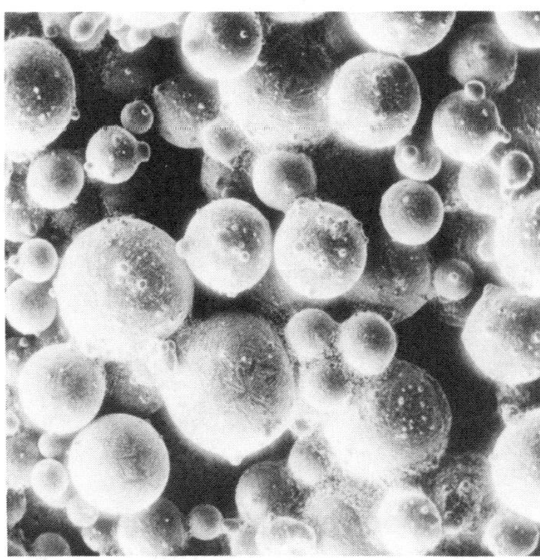

Fig. 10.3 Argon-atomized superalloy powder

argon-gas atomization. In addition to being spherical, large particles often have smaller satellite particles attached to them. Also, particles may have a partial coating of splat, which occurs when a solid particle collides with a liquid particle. Following powder production, powders are screened to remove oversize particles. For superalloys, powder sizes range from -60 mesh (-250 μm) to -325 mesh (-44 μm) depending on the application.

With superalloys, screening is used to limit the maximum nonmetallic inclusion size in the final product. This is a unique feature of the P/M process. After being screened, powders are blended to obtain a uniform size distribution. The powders are then ready for loading into containers for subsequent consolidation.

TYPICAL POWDER CONSOLIDATION PROCESSES

Superalloy powders are consolidated to full density using a combination of temperature and pressure. The two principal techniques used to consolidate superalloy powders for aerospace applications are: (*a*) hot isostatic pressing (HIP), and (*b*) extrusion followed by isothermal forging. For hot isostatic pressing, powder-filled containers are placed in an autoclave that is subsequently heated and pressurized. Superalloys are normally hot isostatically pressed to full density at temperatures ranging from 1095 to 1205 °C (2000 to 2200 °F) under a pressure of 103 MPa (15 ksi).

Hot isostatic pressing can be used to produce parts of simple or very complex shapes depending on the container used for the powder. Metal containers are currently the main type of container used for production of hot isostatic pressing P/M superalloy hardware. These range from simple cylindrical containers made from steel pipe to complex-shape containers made from formed sheet. (See Fig. 10.4.)

A unique ceramic mold process for producing very complex parts has been developed. The mold is produced by the lost-wax process using

Fig. 10.4 Sectioned metal container for a hot isostatically pressed powder multicomponent compressor spool

specially developed materials. The mold is filled with powder and subsequently placed inside a cylindrical metal container that is surrounded by a granular ceramic. The metal container is then sealed. During consolidation, the granular ceramic transmits pressure from the outer container to the ceramic container. The process is capable of producing a variety of very complex shapes. (See Fig. 10.5.)

Extrusion-plus-isothermal forging also is used commercially to consolidate and form superalloy powders into useful shapes. Powders are loaded into cylindrical steel containers, which are evacuated, sealed, and then extruded at high temperature to consolidate the powder and develop a fine, uniform grain structure. This structure is capable of superplastic behavior at elevated temperatures and low strain rates. As such, the material can be formed to relatively complicated shapes using isothermal forging.

Several other techniques are being developed for consolidation of superalloy powders. These include:

* Consolidation by atmospheric pressure (CAP)
* Fluid die process

CONSOLIDATION BY ATMOSPHERIC PRESSURE

In the CAP process, powder is loaded into a glass container that is subsequently evacuated and sealed. The assembly is then heated to a temperature slightly below the liquidus of the alloy being consolidated for 4 to 24 h, depending on the compact size. Consolidation results from the pressure difference between the inside and outside of the container, i.e., ~100 kPa (~15 psi), and from sintering mechanisms. The resulting product is not fully dense and requires subsequent hot isostatic pressing and/or hot working to attain full density and optimum properties.

Jet engine components are candidates for parts processed from hot worked CAP stock. Consolidation by atmospheric pressure is a viable

Courtesy of Crucible Research Center.

Fig. 10.5 Assorted parts made by the ceramic mold process

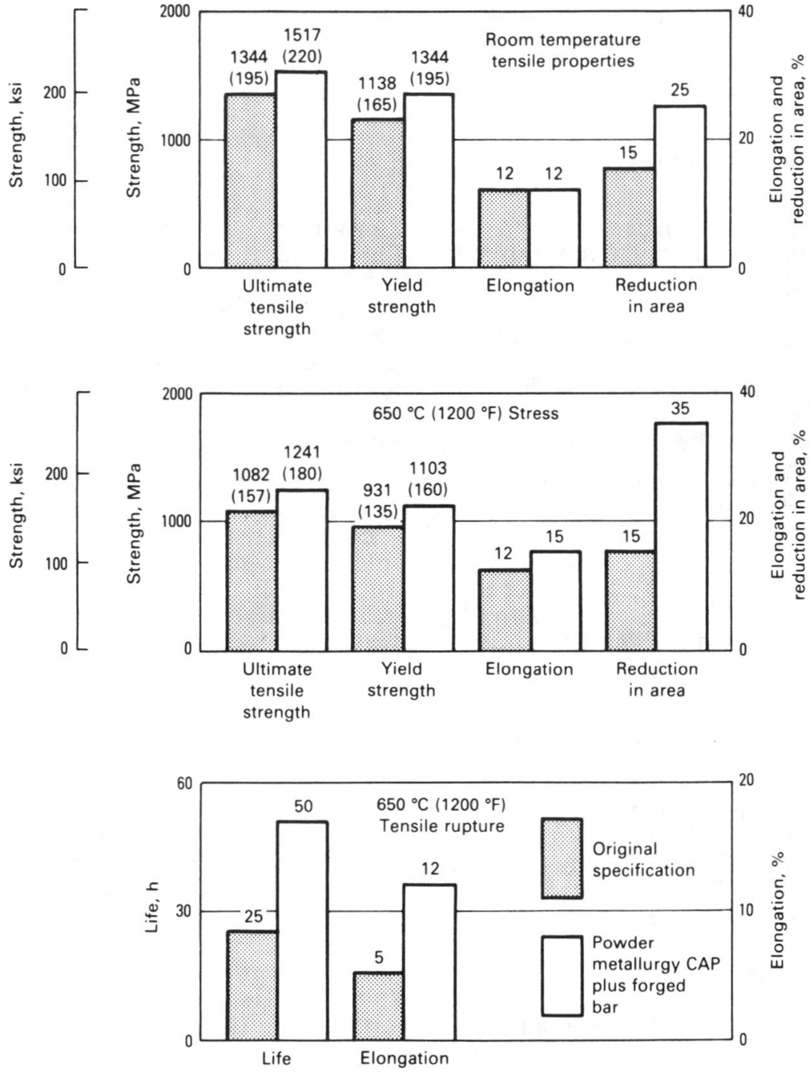

Fig. 10.6 Comparison of mechanical properties of Inconel 718

process for consolidation of superalloy powders such as Rene′ 95 and Inconel 718. Great potential is envisioned for consolidation of rapidly solidified powders. Isothermal forging of CAP billets to near-net shapes offers performance and cost advantages.

Figure 10.6 compares mechanical properties of Inconel 718 produced by the CAP process and subsequent hot forging to original property specifications. Both room-temperature and elevated temperature tensile property requirements and stress-rupture requirements are surpassed by CAP-plus-forged P/M alloys. For direct aged material, there is no sacrifice in ductility at higher strength levels, which indicates metallurgically sound particle-to-particle bonding in the CAP-plus-forged material. For this material, the small amount of boron helps prevent grain-boundary carbide films by promoting the formation of $M_{23}C_6$-type carbides. Also, boron inhibits the formation of cellular orthorhombic delta Ni_3Nb, which tends to form on extended elevated temperature exposure, thus degrading strength.

FLUID DIE PROCESS

The fluid die process uses a shaped cavity that is machined, cast, or forged into mating metal blocks. This process is also referred to as the rapid omnidirectional compaction, or ROC, process. The cavity is filled with powder, evacuated, and sealed. Consolidation is accomplished by applying pressure to the mold by either hot isostatic pressing or forging at high temperature. After consolidation, the mold is removed by machining, leaching, or melting. When melting is used, the metal mold is of a composition that melts at a temperature below that of the P/M superalloy part.

Once consolidated, P/M materials undergo conventional finishing processes such as heat treatment, nondestructive inspection, and machining. In many instances, however, modified heat treatments have been developed for P/M products. Also, machining requirements are often markedly reduced and inspectability of the P/M product is often enhanced. For example, the uniform structure obtained with P/M products is ideally suited to ultrasonic inspection.

FORGING AND ROLLING OF P/M BILLETS

Powder metallurgy techniques can be combined with cast and wrought metalworking technology to produce mill products, billets, or near-net shapes. Such production techniques capitalize on the advantages of powder metallurgy for wrought products. Improvement of performance by producing a wrought product from a P/M billet rather than from an ingot metallurgy billet is the prime reason for acceptance of these processes. Alloys that are subject to heavy segregation, poor castability, poor workability, or that cannot be produced by conventional cast and wrought practice are often candidates for wrought P/M processing.

As the operating temperatures of superalloys in turbine engines have increased, the volume fraction of dispersoids and/or precipitates used for strengthening at elevated temperatures also has increased.

Superalloys capable of withstanding higher operating temperatures have limited workability due to the higher volume fraction of the second phase present. Also, segregation becomes more significant as alloy complexity increases. As a result, conventional ingot metallurgy practices, such as casting followed by forging, are not practical for many current superalloys if optimum properties are required. Primarily, this affects the production of disks and similar turbine engine components that traditionally have been produced by conventional techniques.

Powder metallurgy offers a method of producing a preform shape that may be worked into a desired final shape, thus preventing segregation and workability problems encountered in casting an ingot and working it into a preform shape. For superalloys, conventional press forging or new processes such as isothermal forging can be used for the final working operation. Examples of manufacturing practices for disk production are described below.

Powder is first consolidated to near-theoretical density to form a workpiece. The workpiece is then subjected to a secondary deformation process to produce a mill shape, billet, or near-net shape of theoretical density. Deformation to a final shape is the equivalent of plastic working of a conventional material into an intermediate or final product shape, because the initial workpiece is of near-theoretical density. Selection of the proper thermomechanical treatment depends on the material being processed and the desired final shape. Regardless of the particular process, thermomechanical treatment ensures the desired final shape, complete densification, and massive deformation, which disrupts any prior particle boundaries and contaminants. Powder metals are then processed into wrought P/M products.

Hot Isostatic Pressing Plus Conventional Forging

One approach for producing a workpiece from powder is to hot isostatically press prealloyed powder that has been containerized into a preform shape. These powders are usually spherical, with low oxygen contents; typically, they are produced by inert gas atomization, by soluble-gas atomization, or by the rotating electrode process.

Powder is hermetically sealed in a suitable container and then subjected to an appropriate temperature/pressure/time cycle in an autoclave to produce densification and particle bonding. Microstructure also can be controlled by proper selection of these variables. (See Fig. 10.7.)

Temperatures are selected that promote the formation of intergranular rather than interparticle carbides ($M_{23}C_6$ carbides instead of MC carbides). In practice, densities above 95% and normally above 98% of theoretical are achieved in this preforming step. At this point, the container is removed. The hot isostatically pressed billet may be an individual preform, or it may be sectioned into many preforms for subsequent working into final shapes.

Conventional forging is used next to produce a forged blank that is then machined to final dimensions. Preforms are heated to the forging temperature and forged in one step into a blank. A hydraulic press usually is used because of the size of the part and the reduced flow stress.

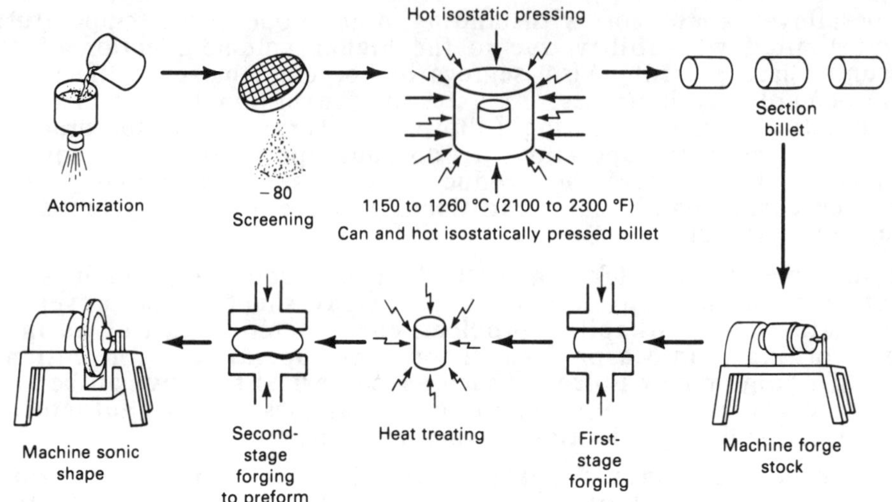

Fig. 10.7 Flowchart of hot isostatic pressing followed by forging for producing turbine disks

Forging temperatures and deformation levels are selected to optimize the mechanical properties of the disk; low-cycle fatigue resistance is of prime importance. Optimum microstructure is a partially warm worked structure, with carbides distributed in a necklace pattern. Full density is ensured by forging, as is disruption of any particle-boundary films.

Hot Isostatic Pressing Plus Isothermal and Hot Die Forging

Hot isostatic pressing plus isothermal and hot die forging are similar to hot isostatic pressing plus conventional forging, except that near-net shape practices of isothermal forging or hot die forging are substituted for conventional press forging. Isothermal forging involves heating the workpiece and tooling to the same temperature and forging at low strain rates. In hot die forging, the workpiece is heated to temperatures above that of the tooling, and the tooling is heated to temperatures well above those of typical forging tooling. Low strain rates are necessary. Sometimes, these processes are referred to as creep deformation processes because of the low strain rates used.

Hot die and isothermal forging processes have greater shapemaking capabilities than conventional press forging for these alloys, due to the use of low strain rates and hot tooling, which eliminates die chill problems. Advantages of these processes over conventional press forging due to low strain rates include:

* Reduced flow stresses in the workpiece
* Enhanced workability
* Greater control over final microstructure
* Greater dimensional precision

The low strain rates capitalize on the fine grain size of P/M preforms for improved workability. These advantages allow production of near-net shapes, which results in significant cost savings due to reduced materials usage and reduced machining. For example, a weight savings of 223 kg (490 lb) of material was realized in the F-100 turbine engine. Conventional press forging of disks required a total of 450 kg (990 lb) of disk material, whereas isothermally forged disks were produced from only 227 kg (500 lb) of starting stock.

Disadvantages of hot isostatic pressing plus isothermal and hot die forging include:

* Long forging cycles (typically 15 min or longer dwell times)
* High cost of forging equipment (atmosphere chambers and integral heating stations)
* Cost of forging dies (expensive alloys such as TZM molybdenum alloy, rather than lower cost tool steels)
* Need for an inert atmosphere or vacuum to protect tooling

Despite these costs, this process combination is economical for producing parts from expensive alloys, where near-net shape production minimizes expensive machining and material losses. Alloys such as Astroloy, IN-100, Rene′ 95, and MERL 76 comprise the bulk of isothermally forged P/M tonnage.

Hot Extrusion Plus Forging

Superalloy powders also can be consolidated into billet shapes by hot extrusion plus forging. Powder, usually containerized, is hot extruded at a reduction ratio of at least 9 to 1 to achieve a fully consolidated billet. The extruded length is then sectioned into workpiece billets for subsequent hot working by conventional press forging, isothermal forging, or hot die forging. For press forging, two steps are required. First, the billet is forged into a preform shape. Then a forged blank is produced from the preform. Preform shaping operations also may be added to the isothermal and hot die forging steps to reduce cycle time in the low-strain rate deformation step. A schematic flowchart of the production of superalloy turbine disks by extrusion followed by isothermal forging is shown in Fig. 10.8.

Mechanical Alloying

Another P/M superalloy use is the mechanical alloying (MA) process developed by INCO Limited. This process is in the early production stage, and a number of alloys manufactured in this manner are available commercially. Mechanical alloying is a dry, high-energy milling process that produces component metal powders with controlled, extremely fine microstructures. The powders are produced in high-energy attrition mills or special large ball mills. The MA products are characterized by oxide dispersion strengthening of superalloy compositions using yttrium oxide as the dispersoid.

Production of a stable powder containing a uniform dispersion of oxide particles is the first step. Subsequently, these powders are

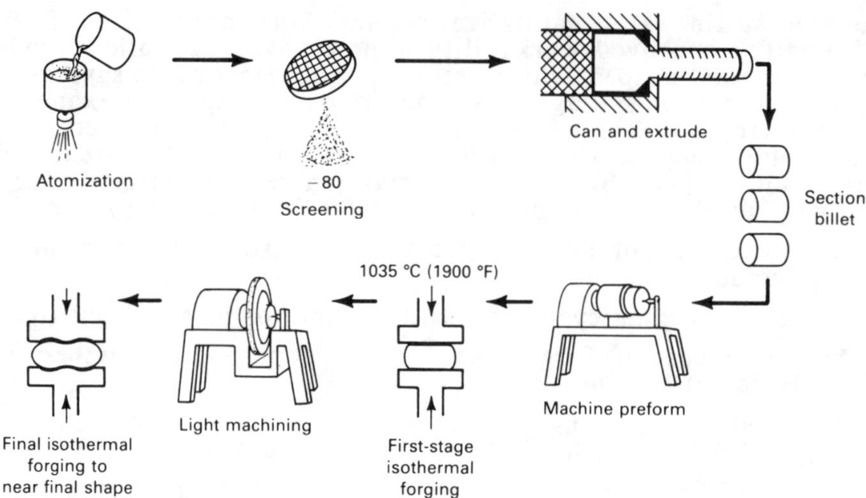

Fig. 10.8 Flowchart of extrusion followed by isothermal forging for production of turbine disks

consolidated by extrusion or hot isostatic pressing. Consolidated pieces are then subjected to hot and/or cold deformation, which imparts a high level of stored energy. Some MA alloys are strengthened by the dispersoid alone, others by the combination and coexistence of the precipitated γ' phase and the dispersed yttrium oxide particle.

For more information on P/M dispersion-strengthened alloys, see "Forging and Rolling of P/M Billets" and the "Dispersion-Strengthened Materials" sections of "Superalloy Powder Metallurgy," *Metals Handbook; Desk Edition.*

MECHANICAL PROPERTIES

Tensile strength, creep resistance, and stress-rupture of consolidated superalloys are comparable to or exceed the conventionally processed ingot material, as shown in Table 10.2 and Fig. 10.9. Fatigue properties of P/M superalloys are comparable to conventional material, although the presence of inclusions in P/M parts produces variations in these properties.

Additional data on the mechanical properties of select superalloys are summarized in Table 10.3. In addition to room-temperature tensile properties, superalloys are characterized by superior elevated temperature properties (stress-rupture and creep) and low-cycle fatigue. In general, the data in Table 10.3 reflect the influence of composition, processing method, and heat treatment on mechanical properties of P/M alloys.

For additional information on mechanical properties of P/M superalloys, see "Powder Systems and Applications" in Volume 7 of the 9th Edition of *Metals Handbook*.

Table 10.2 Comparison of mechanical properties of several P/M superalloys

Condition	Test temperature °C	Test temperature °F	0.2% offset yield strength MPa	0.2% offset yield strength ksi	Ultimate tensile strength MPa	Ultimate tensile strength ksi	Reduction in area, %	Total elongation, %
René 95								
Hot isostatically pressed(a)	23	74	1214	176	1636	237	15	16
Hot isostatically pressed and forged	23	74	1179	171	1629	236	23	18
Cast and wrought	23	74	1144	166	1434	208	12	10
Minimum hot isostatically pressed	650	1202	1120	162	1514	220	17	16
Hot isostatically pressed and forged	650	1202	1122	163	1480	215	14	13
Cast and wrought	650	1202	1055	153	1282	186	10	8
Astroloy								
Hot isostatically pressed	23	74	936	136	1379	200	31	27
Hot isostatically pressed and forged	23	74	1055	153	1517	220	23	27
Hot isostatically pressed	650	1202	881	128	1234	179	36	31
Hot isostatically pressed and forged	650	1202	975	141	1261	183	25	38
IN-100								
Hot isostatically pressed	650	1202	1286	187	942	137	...	21
Hot isostatically pressed and forged	650	1202	1200	174	1000	145	...	8
Hot isostatically pressed and extruded	650	1202	1350	196	1000	145	...	18

(a) 1120 °C (2050 °F) at 103 MPa (15 ksi) for 3 h, solution treated at 1150 °C (2100 °F) 1 h, hot salt quench to 535 °C (1000 °F), aged at 870 °C (1600 °F) 1 h, then 650 °C (1200 °F) 24 h; air cooled

AEROSPACE APPLICATIONS

Although introduction of P/M parts into aerospace systems has been limited because of costly and lengthy qualification requirements, significant P/M part usage has been achieved. For jet engine disk applications (Fig. 10.10), powder superalloy usage is projected to increase to meet the higher mechanical property requirements of future engines.

Despite these P/M advances, however, several problems have been identified with this type of processing, which may affect the long-range potential. These include:

* Metallic and nonmetallic inclusions decrease low-cycle fatigue.
* Understanding the mechanisms responsible for break-up of liquid metal streams via different atomization processes is limited. Correlations among process variables and particle size, shape, and distribution are often empirical.

Table 10.3 Typical mechanical properties of P/M superalloys

Property	René 95(a)	Low-carbon Astroloy(b)	Low-carbon Astroloy(c)	Low-carbon Astroloy(d)	IN-1000(e)	MERL 76(f)	Udimet 700(g)	Udimet 700(a)
0.2% yield strength at 210 °C (410 °F), MPa (ksi)	1257 (182)	973 (141)	928 (135)	994 (143)	1095 (159)	1188 (172)	860 (125)	1115 (162)
Ultimate tensile strength at 210 °C (410 °F), MPa (ksi)	1671 (242)	1376 (200)	1338 (194)	1359 (197)	1594 (231)	1674 (243)	1355 (197)	1515 (219)
Elongation, %	20	22	26	28	26	21	25	18.5
Reduction in area, %	20.3	23	28	32	27	22	27	18.5
Creep at 595 °C (1110 °F) at 1034 MPa (150 ksi), 100 h/% strain	0.15	...	...	...	...	...	...	...
Stress rupture at 650 °C (1200 °F) at 1034 MPa (150 ksi), service life (hours)/% elongation	29.5/5.4	...	...	...	...	...	...	...
	28.4/4.7	...	...	...	...	...	...	...
Strain-controlled low-cycle fatigue at 535 °C (1000 °F), strain/cycles to failure	0.78/26 948	...	...	...	...	...	...	...
	0.66/94 447	...	...	...	...	...	...	...
Stress rupture at 621 MPa (90 ksi) at 730 °C (1350 °F) at 151 h, % elongation/% reduction in area	...	14/17	17/22	16/21	...	...	...	...
0.2% yield strength at								

Property								
					(184)			
% elongation/% reduction in area at 705 °C (1300 °F)	...	...	...	...	19/23	...	...	...
Stress rupture at 730 °C (1350 °F) at 655 MPa (95 ksi), hours to failure/% elongation	...	...	...	...	35.6/16.6	...	...	...
	...	...	...	...	25.5/11.0	...	...	...
	...	...	...	...	37.0/14.5	...	...	...
Creep at 705 °C (1300 °F) at 551 MPa (80 ksi), time for 0.1%/time for 0.2%	...	...	...	...	140.5/193.5	...	...	...
	...	...	...	...	100.0/142.0	...	...	...
	...	...	...	...	91.0/125.0	...	...	...
0.2% yield strength at 620 °C (1150 °F), MPa (ksi)	...	...	...	...	...	1136 (165)	...	...
Ultimate tensile strength at 620 °C (1150 °F), MPa (ksi)	...	...	...	...	...	1492 (216)	...	...
% elongation/% reduction in area at 620 °C (1150 °F)	...	...	...	...	...	18.5/17	...	...

(a) Hot isostatically pressed and hardened and tempered. (b) Produced by rapid omnidirectional compaction. Consolidated at 811 MPa (58.8 tsi); 0.5-s dwell in composite of copper-nickel fluid dies. Preheated to 1075 °C (1970 °F); held at temperature 1 h. Powder was electrodynamically degassed prior to vacuum filling. Post-consolidation solution treated at 1165 °C (2125 °F) for 4 h, fan air cooled. (c) Hot isostatically pressed at 1150 °C (2100 °F) at 104 MPa (15 ksi) for 4 h. Hot loaded at 925 °C (1700 °F). Hot unloaded at 980 °C (1800 °F). Post-consolidation solution treated at 1120 °C (2050 °F) for 2 h, fan air cooled. (d) Hot isostatically pressed and forged. Forging conditions: open die side upset at 1095 °C (2000 °F). Average reduction is 52%. Aging heat treatment for all processes: 650 °C (1200 °F) for 24 h, air cooled plus holding at 760 °C (1400 °F) for 8 h, air cooled. All tensile specimens tested normal to the forging direction. (e) Hot isostatically pressed and gatorized billet. (f) Hot isostatically pressed and gatorized. (g) As hot isostatically pressed

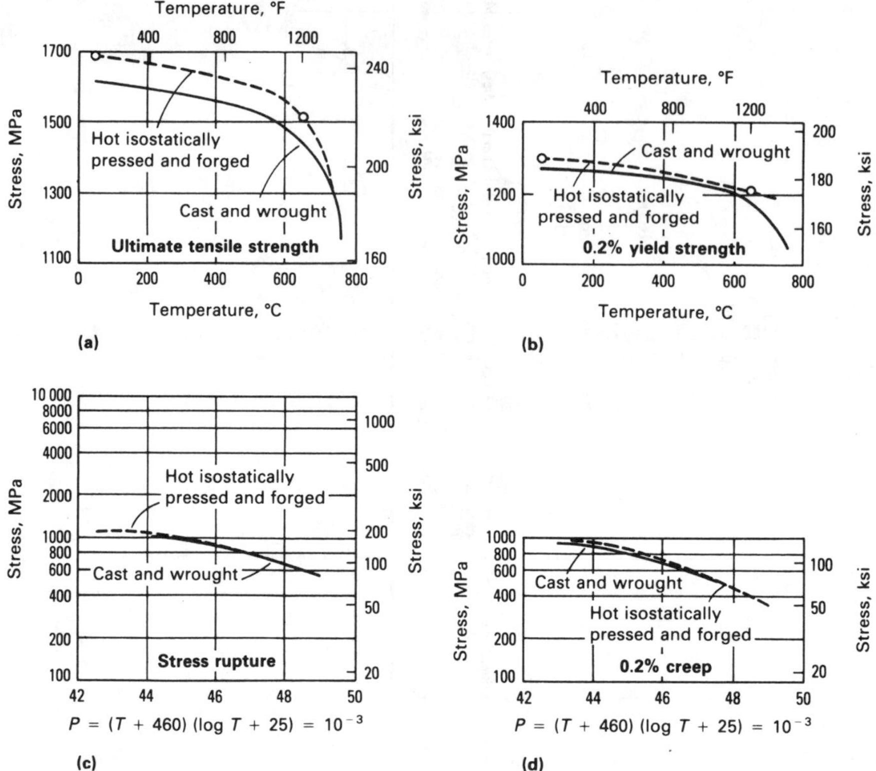

Fig. 10.9 Comparison of tensile and creep properties of hot isostatically pressed P/M and cast and wrought Rene′ 95

* Models for evaluating the role of small flaws on fatigue strength are inadequate.
* Inspection methods for detecting size and number of inclusions or amount of porosity are not fully effective.
* The complex heat-flow and solidification conditions in atomized droplets are not well understood.
* Notwithstanding the near-net shape advantage of P/M processing, component costs are often higher than conventional ingot-forged parts.

P/M materials used for aerospace applications have followed traditional alloy development. P/M superalloys that are currently in volume production for use in jet engine components exhibit high strength at an acceptable cost.

JET ENGINE APPLICATIONS

The demands for improved jet engine efficiency and performance (thrust-to-weight ratio) have resulted in continued development of higher strength superalloys. Concurrent with the increase in strength

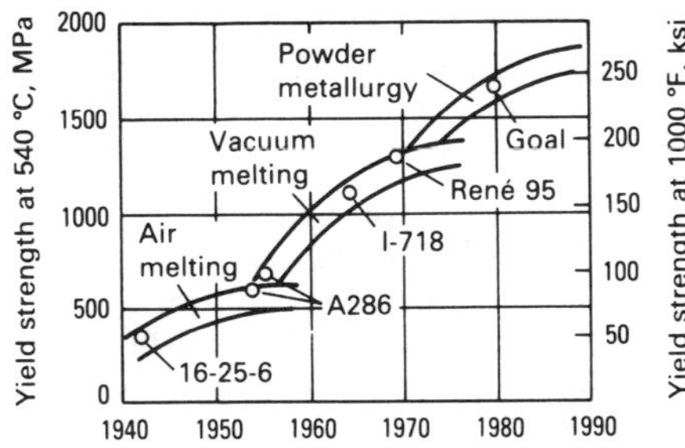

Note growing role of P/M materials as higher strengths at elevated temperatures are required.

Fig. 10.10 Advancement in disk materials

Table 10.4 Current aircraft engine P/M nickel-base superalloys

	Composition, %										
Alloy	Cr	Co	Mo	Nb	V	Ti	Al	Zr	B	Hf	C
Astroloy	15	17	5	...	...	3.5	4	0.05	0.03	...	0.06
IN-100	12.4	18.5	3.2	...	0.8	4.3	5	...	0.02	...	0.07
Rene' 95	14	8	3.5	...	...	2.5	3.5	0.05	0.01	...	0.05
MERL 76	12.5	18.5	3.2	1.35	...	4.4	5	...	0.02	0.4	0.025

exhibited by these superalloys is a significant decrease in conventional hot workability. Reduction in hot workability adversely affects overall component cost. P/M techniques offer a means of producing these highly alloyed compositions with a uniform and homogeneous microstructure that is superplastic and, therefore, suitable for hot working. This has resulted in the replacement of cast and wrought materials by P/M counterparts in many high-performance applications. See Table 10.4 for current aircraft engine P/M superalloys.

A number of case histories involving various types of superalloys as applied to jet engine parts may be found in "Aerospace Applications" in Volume 7 of the 9th Edition of *Metals Handbook.*

DUAL-ALLOY/DUAL-PROPERTY ENGINE COMPONENTS

Powder metallurgy technology is currently under development to produce dual-alloy, dual-property integral turbine disks. By use of this technology, the high tensile and low-cycle fatigue property requirements of the bore can be attained with a fine-grained P/M product. Creep and stress-rupture requirements of the rim and/or blades can be attained by special processing techniques, such as a single-crystal casting, directionally solidified casting, fiber reinforcement, or mechanical alloying. The ceramic mold process is being developed to produce dual-property turbine wheels for small engines.

One concept uses a blade ring made by tack welding individually cast single-crystal blades. The ring is then sandwiched between two halves of a ceramic mold. The mold is then filled with prealloyed superalloy powder, and the assembly is hot isostatically pressed to consolidate the powder and bond it to the blades. Figure 10.11 shows turbine wheels made by this technique. Figure 10.12 shows the microstructure of the bond line between the powder and cast materials.

Another dual-property wheel technique involves the use of a preconsolidated hot isostatically pressed P/M hub and a directionally solidified cast blade ring. The two components are joined by hot isostatic pressing. Figure 10.13 shows a wheel manufactured by this process. As illustrated, the coarse-grained directionally solidified rim and the fine-grained P/M hub are clearly visible.

Fig. 10.11 Dual-property turbine wheel made from cast blades and a P/M hub

Fig. 10.12 Microstructure at hot isostatically pressed bond line between cast alloy C103 (top) and P/M alloy PA-101 (bottom)

Courtesy of Detroit Diesel Allison.

Fig. 10.13 Dual-property turbine wheel made by bonding (hot isostatically pressing) a fine-grained hub to a coarse-grained directionally solidified cast blade ring

Chapter 11

Heat Treating

INTRODUCTION

As with other alloys, superalloys sometimes require heat treatment. Some superalloy compositions are hardened and strengthened by heat treatment, and at other times, stress relieving or annealing treatments are required. Typical thermal treatments are discussed in this chapter. Table 3.1 in Chapter 3 lists the important superalloys by various classes: wrought, cast, nickel-base, iron-base, and cobalt-base alloys. For more detailed information on this subject, see "Heat Treating of Heat-Resisting Alloys" in Volume 4 of the 9th Edition of *Metals Handbook.*

STRESS RELIEVING

Stress relieving of superalloys frequently entails a compromise: the desirability of maximum relief of residual stress must be weighed against possible deleterious effects to high-temperature properties and corrosion resistance. True stress relieving of wrought material usually is confined to alloys that are not precipitation strengthening. Thus, the time and temperature cycles may vary considerably, depending on the metallurgical characteristics of the alloy and on the type and magnitude of residual stresses developed by previous fabricating processes.

Stress relieving temperatures are usually below the annealing or recrystallization temperatures. Typical cycles for wrought alloys are listed in Table 11.1; temperatures at least 25 °C (50 °F) higher or lower than those listed are usually satisfactory.

Some superalloy castings are placed in service in the as-cast condition. However, castings may be stress relieved: (*a*) when they are of an extremely complex shape that might crack during the initial heating-up period in service; (*b*) when their dimensional tolerances are stringent; and (*c*) after they have been welded.

It is not possible to tabulate the stress relief cycles for cast alloys, because they are particularly dependent on geometry and prior processing. Stress relief cycles can be developed for many alloys by empirical studies of stress decay with time and temperature, as determined by nondestructive means such as X-ray diffraction. This is not an effective technique for superalloys, where extensive material testing of critical properties and subsequent data analyses must be performed to determine the efficacy of

Table 11.1 Typical stress relieving and annealing cycles for wrought superalloys

Alloy	Stress relieving: Temperature °C	Stress relieving: Temperature °F	Stress relieving: Holding time per inch of section, h	Annealing(a): Temperature °C	Annealing(a): Temperature °F	Annealing(a): Holding time per inch of section, h
Iron-base and iron-nickel-chromium alloys						
RA-330	900	1650	1(b)	1110(c)	2025(c)	1/4(d)
A-286	(f)	(f)	...	980	1800	1
Discaloy	(f)	(f)	...	1035	1900	1
Nickel-base alloys						
Astroloy	(f)	(f)	...	1135	2075	4
Hastelloy B	(f)	(f)	...	1175	2150	1
Hastelloy C	(f)	(f)	...	1215	2225	1
Hastelloy W	(f)	(f)	...	1175	2150	1
Hastelloy X	(f)	(f)	...	1175	2150	1
Incoloy 800	870	1600	1 1/2	980	1800	1/4
Incoloy 800H	...	...	...	1175	2150	...
Incoloy 825	...	...	...	980	1800	...
Incoloy 901	(f)	(f)	...	1095	2000	2
Inconel 600	900	1650	1	1010	1850	1/4(d)
Inconel 601	...	...	...	980	1800	...
Inconel 625	870	1600	1	980	1800	1
Inconel 690	...	...	...	1040	1900	1/2

René 41	(f)	(f)	...	1080	1975	2
Udimet 500	(f)	(f)	...	1080	1975	4
Udimet 700	(f)	(f)	...	1135	2075	4
Waspaloy	(f)	(f)	...	1010	1850	4
Cobalt-chromium-nickel-base alloys						
L-605 (HS-25)	(h)	(h)	...	1230	2250	1
N-155 (HS-95)	(h)	(h)	...	1175	2150	...
S-816	(h)	(h)	...	1205	2200	1
Refractory metals(j)						
Ta-10W	1205(k)	2200(k)	1	1425(k)	2600(k)	1
FS-80	1095(k)	2000(k)	1	1315(k)	2400(k)	1
FS-82	1095(k)	2000(k)	1	1315(k)	2400(k)	1
Mo-0.5 Ti	1095(m)	2000(m)	1/2	1315(m)(n)	2400(m)(n)	1
TZM	1205(m)	2200(m)	1	1425(m)(n)(p)	2600(m)(n)(p)	1

(a) Minimum hardness is achieved by cooling rapidly from the annealing temperature, to prevent precipitation strengthening of phases. Water quenching is preferred and is usually necessary for heavy sections; air cooling is preferred for heavy sections of Waspaloy, Udimet 500, Udimet 700, and Inconel X-750, because water quenching causes cracking. However, for complex shapes subject to excessive distortion, oil quenching is often adequate and more practical. Rapid air cooling usually is adequate for parts formed from strip or sheet. Rapid cooling from the annealing or solution treating temperature does not suppress the aging reaction of some alloys, such as Astroloy; these alloys become harder and stronger. (b) Time given is minimum; some plants use as long as 3 h per inch. (c) Nominal temperature; 1035 to 1175 °C (1900 to 2150 °F) is commonly used. (d) Short time is required for prevention of grain coarsening. (e) Nominal temperature; 650 to 705 °C (1200 to 1300 °F) is permissible. (f) Full annealing is recommended, because intermediate temperatures cause aging. (g) Used only for stress equalizing of warm worked grades. (h) Full annealing is recommended if further fabrication is performed; otherwise, material can be stress relieved at approximately 55 °C (100 °F) below annealing temperature. (j) Annealing temperatures depend on prior plastic deformation, degree of cold work, alloy content, and interstitial purity. Annealing temperatures given are those most frequently used for cold worked sheet or plate; in many instances, more precise determination of the recrystallization temperature is necessary for a specific application.

a given cycle. Particular care must be given to such time-dependent effects as low-cycle fatigue (LCF) hold time, crack growth, and creep rate.

ANNEALING

When applied to superalloys, annealing implies full annealing - that is, complete recrystallization and the attainment of maximum softness. This practice usually is applied to wrought alloys of the non-precipitation-strengthening type. For a majority of the strengthenable alloys, annealing cycles are the same as those used for solution treating. However, the two treatments serve different purposes. Annealing is used mainly: to increase ductility (and reduce hardness) to facilitate forming or machining; to prepare for welding; to relieve stresses after welding; to produce specific microstructures; or to soften age-hardened structures by re-solution of second phases.

By contrast, solution treating is intended to dissolve second phases to produce maximum corrosion resistance, or to prepare for aging treatments at intermediate temperatures for strengthening purposes. Additionally, it will homogenize microstructure prior to aging.

Annealing practices vary considerably among different plants. Representative annealing temperatures, holding times, and cooling procedures are given in Table 11.1. Experience with specific parts for known requirements often indicates advantageous modifications of temperature, hold time, or cooling method.

Most wrought superalloys can be cold formed, but they are more difficult to form than austenitic stainless steels. Severe cold forming may require several intermediate annealing operations. Full annealing must be followed by rapid cooling.

Annealing of weldments should immediately follow welding of the precipitation-strengthened alloys where highly restrained joints are involved. If the configuration of the weldment does not permit annealing, aging can be used for stress relieving the joints.

Reheating for hot working is an annealing practice that promotes adequate formability of the metal being deformed. Temperatures vary widely depending on alloy and working practice. Control of temperature can be critical to resultant properties, as varying degrees of recrystallization may be desired. In most standard operations, heating or reheating for hot working is a full annealing step with recrystallization and dissolution of all or most secondary phases. Occasionally, reheating for hot working is restricted to temperatures that do not dissolve all secondary phases so that the remaining phases can be used to limit grain growth.

QUENCHING

Quenching of superalloys is performed to retain, at room temperature, the solid solution obtained during solution treating. Because more strengthening phase is in solution at room temperature than is normally possible, the quench produces a supersaturated solution. Quenching also permits a finer gamma prime (γ') particle size to be achieved on aging. Cooling methods commonly used are indicated in Table 11.2 for wrought alloys and in Table 11.3 for casting alloys.

AGING

Aging treatments strengthen superalloys by causing the precipitation of one or more phases from the supersaturated solid-solution matrix that is developed by solution treating and retained by rapid cooling from the solution treating temperature. Factors that influence the selection of the number of aging steps and aging temperature include: (*a*) type and number of precipitating phases available; (*b*) anticipated service temperature; (*c*) precipitate size; (*d*) the combination of strength and ductility desired; and (*e*) the heat treatment of similar alloys.

Principal precipitated phases in superalloys usually include one or more of the following: γ' Ni_3Al or $Ni_3(Al,Ti)$, eta (η, Ni_3Ti), or γ'', body-centered tetragonal (bct) Ni_3Nb. Secondary phases that may be present include carbides ($M_{23}C_6$, M_7C_3, M_6C, and MC), nitrides (MN), carbonitrides (MCN), and borides (M_3B_2), as well as Laves phase (M_2Ti) and delta phase (δ, orthorhombic Ni_3Nb). These phases occur principally in nickel-base alloys. The primary phases in cobalt-base alloys are $M_{23}C_6$, M_7C_3, M_6C, and MC. The primary phases in iron-base alloys are similar to those in nickel alloys, although η is more apt to be present because of the generally higher titanium-aluminum ratio found in iron alloys than in nickel alloys.

When more than one phase is capable of precipitating from the alloy matrix, judicious selection of a single aging temperature may result in obtaining optimum amounts of multiple precipitating phases. Alternatively, a double aging treatment that produces different sizes and types of precipitate at different temperatures may be employed. The aging temperature also determines not only the type, but also the size distribution of precipitate.

HEAT TREATMENT FOR HIGH STRENGTH

Strengthening of superalloys usually requires solution treating and aging. Typical cycles for wrought alloys are given in Table 11.2 and in Table 11.3 for casting alloys. Note that the cooling rate from the solution temperature is critical for some alloys. Alternate heat treatments may be used to improve specific properties.

Solution Treating

In some instances, the solution treating temperature employed will depend on the properties desired, as indicated in Table 11.2 for A-286, Inconel 718, Rene′ 41, Udimet 700, and Waspaloy. A higher temperature is specified for optimum creep and creep-rupture properties, whereas a lower temperature is required for optimum short-time tensile properties at elevated temperature. The higher solution treating temperature will result in some grain growth and more extensive solution of carbides in wrought alloys. The principal objective is to put γ'-type phases into solution and dissolve some carbides.

Table 11.2 Typical solution treating and aging cycles for wrought superalloys

Alloy	Solution treating Temperature °C	Solution treating Temperature °F	Solution treating Time, h	Solution treating Cooling procedure	Aging Temperature °C	Aging Temperature °F	Aging Time, h	Aging Cooling procedure
Iron-base alloys								
A-286	980	1800	1	Oil quench	720	1325	16	Air cool
Discaloy	1010	1850	2	Oil quench	730	1350	20	Air cool
					650	1200	20	Air cool
N-155	1175	2150	1	Water quench	815	1500	4	Air cool
Nickel-base alloys								
Astroloy	1175	2150	4	Air cool	845	1550	24	Air cool
	1080	1975	4	Air cool	760	1400	16	Air cool
Hastelloy B	1175	2150	½	(a)	(b)	(b)	...	...
Hastelloy B-2	1065	1950	½	Rapid quench	...	...	...	...
Hastelloy C-4	1065	1950	½	Rapid quench	...	...	...	...
Hastelloy C-276	1120	2050	½	Rapid quench	...	...	...	...
Hastelloy N	1175	2150	½	Rapid quench	...	...	...	...
Hastelloy S	1065	1950	½	Rapid quench	...	...	...	...
Hastelloy C	1220	2225	1	(a)	(b)	(b)	...	...
Hastelloy W	1175	2150	1	(a)	(b)	(b)	...	...
Hastelloy X	1175	2150	1	(a)	...	...	...	...
Inconel 901	1095	2000	2	Water quench	790	1450	2	Air cool
					720	1325	24	Air cool
Inconel 600	1120	2050	2	Air cool	...	...	...	...
Inconel 601	1150	2100	1	Air cool	...	...	...	...
Inconel 617	1175	2150	2	(a)	...	...	...	...
Inconel 625	1150	2100	2	(a)	...	...	...	...
Inconel 706	925-1010	1700-1850	...	...	845	1550	3	Air cool
					720	1325	8	Furnace cool
					620	1150	8	Air cool

Inconel X-750 (AMS 5667)	855	1625	24	Air cool	705	1300	20	Air cool
Inconel X-750 (AMS 5668)	1150	2100	2	Air cool	845	1550	24	Air cool
					705	1300	20	Air cool
Nimonic 80A	1080	1975	8	Air cool	705	1300	16	Air cool
Nimonic 90	1080	1975	8	Air cool	705	1300	16	Air cool
René 41	1065	1950	1/2	Air cool	760	1400	16	Air cool
Udimet 500	1080	1975	4	Air cool	845	1550	24	Air cool
					760	1400	16	Air cool
Udimet 700	1175	2150	4	Air cool	845	1550	24	Air cool
	1080	1975	4	Air cool	760	1400	16	Air cool
Waspaloy	1080	1975	4	Air cool	845	1550	24	Air cool
					760	1400	16	Air cool
Cobalt-base alloys								
Haynes 25; L-605	1230	2250	1	Rapid air cool	(b)	(b)	...	...
Haynes 188	1175	2150	1/2	Rapid air cool	...	...	...	...
Haynes 556	1175	2150	1/2	Rapid air cool	...	...	...	...
S-816	1175	2150	1	(a)	760	1400	12	Air cool
Stellite 6B	1230	2250	1	Air cool	...	...	...	...

Note: Alternate treatments may be used to improve specific properties. (a) To provide an adequate quench after solution treating, it is necessary to cool below about 540 °C (1000 °F) rapidly enough to prevent precipitation in the intermediate temperature range. For sheet metal parts of most alloys, rapid air cooling will suffice. Oil or water quenching is frequently required for heavier sections that are not subject to cracking. (b) Aging occurs in service at elevated temperatures.

Table 11.3 Typical solution treating and aging cycles for cast superalloys

Alloy	Solution treating: Temperature(a) °C	°F	Time, h	Cooling procedure	Aging: Temperature(b) °C	°F	Time, h	Cooling procedure
A-286	1095	2000	2	Rapid cool	720	1325	16	Air cool
B-1900	As cast		...	...	...	...	...	...
FSX-414	1150	2100	4	Rapid cool	980	1800	4	Air cool
Hastelloy B	1175	2150	2	Rapid cool	(c)	(c)	...	...
Hastelloy C	1220	2225	1	Rapid cool	(c)	(c)	...	...
HS-31 (X-40)	As cast		...	...	...	...	...	...
IN-100	As cast		...	...	...	...	...	...
IN-713C	As cast		...	...	...	...	...	...
IN-738	1120	2050	2	Air cool	845	1550	24	Air cool
IN-792	1120	2050	2	Air cool	845	1550	24	Air cool
IN-939	1160	2120	4	Air cool	850	1560	16	Air cool
Inconel 718	1095	2000	1	Air cool	620	1150	10	Air cool
MAR-M 200	...	...	...	...	870	1600	50	Air cool
MAR-M 200 DS	1230	2250	4	Air cool	870	1600	32	Air cool
MAR-M 246	...	...	...	...	845	1550	50	Air cool
MAR-M 247	...	...	...	...	870	1600	16	Air cool
MAR-M 302	As cast		...	...	...	...	...	...
MAR-M 509	As cast		...	...	...	...	...	...
René 41	1095	2000	1/2	Rapid cool	900	1650	4	Air cool
René 80	1220	2225	2	Air cool	1095	2000	4	Air cool
					1055	1925	4	Air cool
					845	1550	16	Air cool
Udimet 700	1150	2100	2	Air cool	760	1400	16	Air cool

(a) Furnace temperature tolerance of ±15 °C (±25 °F) is satisfactory. (b) Furnace temperature tolerance of ±10 °C (±15 °F) is recommended. (c) Aging occurs in service at elevated temperature. Use a vacuum or protective atmosphere for heat treating at temperatures above 1040 °C (1900 °F) and subsequent cooling.

After aging, the resulting microstructure of these wrought alloys consists of large grains that contain the principal strengthening phases and a heavy concentration of carbides in the grain boundaries. The lower solution treating temperature dissolves the principal strengthening phases without grain growth or significant high-temperature carbide solution.

For some wrought alloys, for example, Nimonic 80A and Nimonic 90 (Table 11.2), an intermediate solution treating temperature is selected to produce a compromise of properties. For other alloys, such as Udimet 500 and Udimet 700, the intermediate-temperature aging treatment is used to modify the grain boundaries for improved creep-rupture properties.

Exposure to temperatures higher than the optimum aging temperature results in a decrease in strength through the process of overaging; at still higher temperatures, re-solution may occur. Aging temperatures higher than optimum produce coarser γ' particles than lower temperatures and result in lower creep-rupture strength. For optimum retention of low-to-intermediate temperature tensile strength achieved through strain hardening (thermomechanical processing), lower final aging temperatures are employed than those used to obtain high creep-rupture properties. For all γ' precipitation, care must be taken to ensure the correct carbide

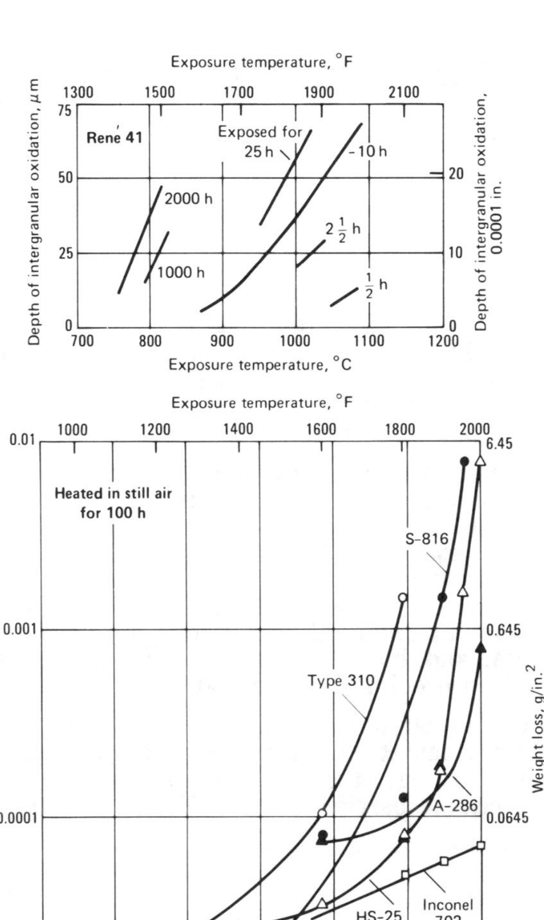

(a) Rene′ 41. (b) Type 310 stainless steel and typical heat-resistant alloys.

Fig. 11.1 Effect of time and temperature on oxidation

distribution. A principal reason for two-step aging sequences, in addition to γ' or γ'' control, is the need to precipitate or control grain-boundary carbide morphology.

SURFACE CONTAMINATION

Some precipitation-strengthened superalloys have good oxidation resistance within their normal range of service at 760 to 980 °C (1400 to 1800 °F). This operating temperature may be at, or slightly above, the aging temperature used for a given alloy. Other superalloys, particularly the high-strength, high-temperature alloys with chromium contents as low as

6 to 8%, may require coatings to provide satisfactory oxidation resistance. Examples of parts that require coatings are turbojet engine turbine airfoils. However, at temperatures used for solution treating, these alloys are susceptible to intergranular oxidation, a defect that adversely affects thermal fatigue.

Intergranular Oxidation

Intergranular oxidation is measured optically as depth of intergranular penetration. Figure 11.1(a) shows the depth of intergranular oxidation that occurred in Rene' 41 that was heated in air. Oxidation data as determined by weight change for several precipitation-strengthened superalloys and type 310 stainless steel are plotted in Fig. 11.1(b).

Oxidation resistance is enhanced by chromium and aluminum additions, as well as by certain other elements. Chromium allows the formation of a protective chromium oxide and is most beneficial at intermediate temperatures and below 870 °C (1600 °F), whereas aluminum forms aluminum oxide, which provides more protection at the higher temperature above 870 °C (1600 °F). The principal mode of intergranular attack involves the preferential oxidation of chromium, aluminum, titanium, zirconium, and boron. Molybdenum increases susceptibility to intergranular attack in precipitation-strengthened superalloys.

In relation to intergranular oxidation, aluminum is preferable to titanium as a strengthening element for two reasons: (*a*) increased aluminum reduces formation of η phase from γ', and (*b*) aluminum oxide provides a denser and less permeable barrier to the diffusion of oxygen.

Carbon Pickup

Carbon pickup can occur if the solution treating atmosphere has a carburizing potential. For instance, the carbon content of the surface of A-286 alloy has been observed to increase from 0.05 to 0.30%. The added carbon forms a stable carbide (TiC), thus removing titanium from solid solution and preventing normal precipitation strengthening in the surface layers. TiN can be formed in the same manner, with the same deleterious effect, as a result of nitrogen contamination.

Miscellaneous Contaminants

All exposed surfaces of superalloy parts should be kept free of dirt, fingerprints, oil, grease, forming compounds, lubricants, and scale. Lubricants or fuel oils that contain sulfur-bearing compounds are particularly active in corroding the metal surface by first forming Cr_2S_3 and then, as the attack progresses, also forming a $Ni\text{-}Ni_3S_2$ eutectic that melts at 645 °C (1190 °F), particularly at low pressures of less than 10^{-4} torr in vacuum.

Scale and slag from furnace hearths are another source of contamination. Contact with steel scale, slag, and furnace spallings should be avoided. Low-melting constituents can form on the metal surface and thus promote corrosion.

ATMOSPHERES

Protective Atmospheres

Protective atmospheres are used in annealing or solution treating if heavy oxidation cannot be tolerated. If oxidation can be tolerated, because of subsequent stock removal, superalloys can be solution treated in air or in the normal mixture of air and combustion products found in gas-fired furnaces.

Typical types of atmospheres include:

* Exothermic
* Endothermic
* Dry hydrogen
* Dry argon
* Vacuum

Atmospheres for Aging

Air is the most common aging atmosphere. The smooth, tight oxide layer that is formed is usually not objectionable on the finished product. However, if this oxide layer must be minimized, a lean exothermic gas with an air-gas ratio of about 10-to-1 can be employed. It will not entirely prevent oxidation, but the oxide layer will be very light. *The use of gases containing hydrogen and carbon monoxide for aging cycles is dangerous because of the explosion hazard at temperatures below 760 °C (1400 °F).*

RELATIONSHIP OF FABRICATION TO HEAT TREATING

A uniformly fine-grain microstructure can be produced in precipitation-strengthened superalloys if final deformation is carried out in the lower part of the hot working range. This type of microstructure can be solution treated to a uniform grain size. If the final working temperature is too low, or if the metal is not reduced enough during final working, a duplex structure is produced during solution treating. Alloys that are deformed in the upper hot working range have a coarser structure that cannot be refined by solution treating.

Cold Working

Cold working usually is performed on alloys in the solution treated condition because of the markedly lower strength and increased ductility of the material before aging. (See Table 11.4.) Cold working affects mechanical properties through its influence on: (*a*) grain growth during subsequent solution treatment, and (*b*) the reaction kinetics of aging.

Table 11.4 Typical effects of aging on room-temperature mechanical properties of solution treated superalloys

Alloy	Yield strength(a) Not aged MPa	ksi	Aged MPa	ksi	Elongation in 50 mm (2 in.), % Not aged	Aged
A-286	240	35	760	110	52	33
René 41	620	90	1100	160	45	15
Inconel X-750	410	60	650	92	45	24
HS-25	480	69	480	70	55	45

(a) At 0.2% offset

Effect of Grain Size

Precipitation-strengthened superalloys are susceptible to critical grain growth if they are solution treated after small amounts of cold or hot work. Larger amounts of cold work refine the grain size. Excessive grain growth may have deleterious effects on fatigue properties in all superalloys.

Therefore, on parts subjected to cold or hot work prior to solution treating, the critical amount of work (about 1 to 6% cold work, depending on the alloy, or 10% hot work) must be exceeded in all areas, to avoid the growth of abnormally large grains. This rule applies to items such as cold headed bolts, spun or stretch-formed sheet, and parts formed by simple bending.

Effect on Aging

Cold working accelerates the aging reaction, causing the early appearance of precipitates at normal aging temperatures, as well as the appearance of a precipitate at temperatures below the normal aging temperature, if the alloy is cold worked sufficiently and held at temperature long enough. In other words, cold working renders the material more susceptible to overaging at the normal aging temperature. Properly charted information correlating cold work with aging response, however, can be used to shorten the required aging period or to lower the normal aging temperature.

Rates of Work Hardening

In cold forming operations such as sheet rolling, tube drawing, pressing, brake and stretch forming, and deep drawing, the governing material property is the rate of hardening in relation to the amount of cold deformation. Materials such as A-286, Nimonic 75, and Hastelloy X can be reduced as much as 90% before a softening treatment is required. For the more complex alloys such as Rene′ 41 and Udimet 500, incremental reductions of 35 to 40% can be achieved without intermediate annealing.

Figure 11.2 shows the relative work-hardening rates of four superalloys compared to type 302 stainless steel. The iron-base A-286 alloy work hardens at about the same rate as type 302. The highest work-hardening rates are those of cobalt-base alloys, such as S-816.

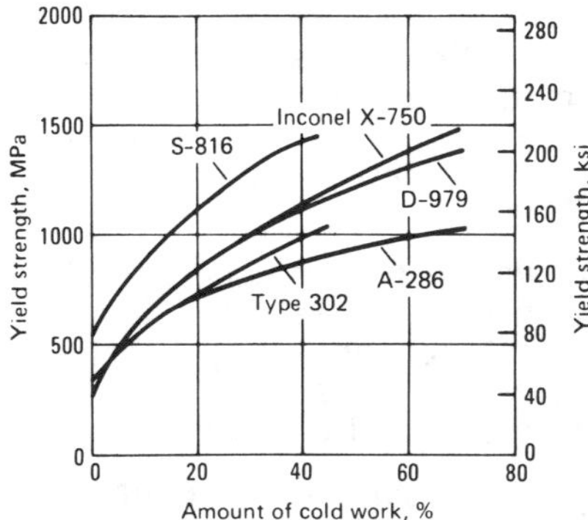

Fig. 11.2 Effect of cold work on room-temperature yield strength of some heat-resistant alloys

In general, the greater the work-hardening rate of a given alloy, the less the amount of cold deformation possible before annealing is required. When a large amount of total reduction is required, as in the draw swaging of tubes, the metal is formed in small incremental reductions with interstage anneals between successive reductions.

Welding

Welding of precipitation-strengthened superalloys requires precise control of prewelding and postwelding heat treatments. A typical welding sequence for wrought alloys is as follows:

* Material received in mill annealed condition (The specific condition is critical for certain alloys.)
* Forming by various methods (in-process anneals used if required)
* Welding
* Solution treating of weldment
* Aging of weldment

Mill annealing cycles must dissolve precipitated phases without causing excessive grain growth and must not dissolve M_6C in alloys such as Rene′ 41. Large grain size and dissolved carbides may produce a continuous grain-boundary carbide precipitate that promotes cracking during welding. A maximum grain size of ASTM 3 usually is recommended for welding.

Fusion welds in precipitation-strengthened superalloys should be solution treated after welding to eliminate the uncontrolled aging that occurs in the heat-affected zone of the base metal during welding. Solution treatment should also be performed to relieve residual stresses. Special procedures, such as double aging, have been used in large and complex components containing restrained welds, for which solution treatment is not practical.

Cast non-precipitation-strengthened cobalt-base superalloy turbine vanes are repair welded, and heat treatment procedures, such as preheat and post-heat, have been developed to minimize cracking and surface contamination in these alloys. Precipitation-strengthened cast nickel-base superalloys normally are not repair welded, with the exception of the weldable superalloy Inconel 718. Cracking and dimensional control problems commonly are encountered with the welding of cast nickel-base γ'-strengthened superalloys.

Service requirements of weldments must be considered in the selection of heat treatment. If service temperatures do not exceed the aging temperature, no appreciable change in properties is likely in service. However, if the service temperature is above the normal aging temperature, strength decreases and ductility increases in service.

Heat treatments that may be applied to weldments in solution-strengthened superalloys are limited to stress relieving and, in some instances, stabilizing. Stress relieving must be done at relatively high temperatures of 815 to 980 °C (1500 to 1800 °F) to be effective. Because the stress-relieving temperature range coincides with the carbide precipitation range, the composition of the alloys should be controlled to avoid precipitating carbides in harmful amounts.

HEAT TREATING OF IRON-BASE ALLOYS

Iron-base superalloys include some alloys that are precipitation strengthened by solution treating and aging and some that are not strengthened by heat treatment. Heat treating temperatures for select alloys are given in Tables 11.1, 11.2, and 11.3.

Stress Relieving and Annealing

Alloys that depend on hot and cold work to augment solid-solution strengthening frequently require stress relieving (Table 11.1). However, these alloys cannot be stress relieved when maximum corrosion resistance is required in service, because the recommended stress relieving temperatures are within the sensitizing range of these alloys. If these alloys are to be exposed to acid or steam in service, intergranular cracking can be minimized, or eliminated, by annealing instead of stress relieving.

Iron-base superalloys, such as A-286 and Incoloy 901, which rely on γ' precipitation strengthening, although not susceptible to sensitization at stress relieving temperatures, still cannot be stress relieved, because the intermediate temperatures of stress relief result in precipitation. Consequently, the restoration of ductility and reduction of stresses in cold formed parts and weldments made of these alloys is achieved by heating rapidly to the annealing temperature. This procedure will result in partial solutioning of the γ' phase, reduction in residual stress, and subsequent reprecipitation of γ' on cooling. In forging these alloys, the finishing temperature is usually above 925 °C (1700 °F), and stress relieving is not required for as-forged parts. These alloys usually are solution treated and aged after forging.

Castings

Iron-chromium-nickel alloys, identified by the Alloy Casting Institute designations HA through HX, usually are exposed to service temperatures of 650 to 1250 °C (1200 to 2200 °F) in process equipment such as heat treating furnaces, oil refinery equipment, and cement kilns. Some of these compositions, such as HA and HF, are more correctly identified as stainless steels rather than superalloys. Normally, the iron-chromium-nickel alloys are placed in service in the as-cast condition, but when experience indicates the advisability of stress relieving, the operation is performed by placing the castings in a cold furnace and heating slowly to 980 to 1040 °C (1800 to 1900 °F), holding at temperature for 1 h per inch of section thickness, and then furnace cooling.

Castings usually are not annealed, because the need to restore ductility seldom exists, and slow cooling through the temperature range of 705 to 900 °C (1300 to 1650 °F) may cause formation of the brittle CrFe σ phase in some grades, such as HD where chromium content approaches 30%. Therefore, these castings are annealed only for special reasons, as discussed below. However, in some instances, stress relieving operations are performed to relieve residual stresses.

Iron-chromium casting alloys containing 9 to 30% chromium and little or no nickel vary widely in response to heat treatment, depending on composition. Most castings of alloys in this group are used in the as-cast condition. Type HA (9% Cr, 1% Mo) is an exception, however; it transforms from austenite to ferrite on air cooling from temperatures above 815 °C (1500 °F). For maximum softness, castings of this alloy are annealed by furnace cooling from 885 °C (1625 °F) minimum at about 25 °C/h (50 °F/h) to below 705 °C (1300 °F). As indicated previously, HA and similar compositions are actually stainless steels.

Iron-chromium-nickel and iron-nickel-chromium casting alloys containing 20 to 30% Cr and 10 to 20% Ni (such as HE, HF, and HL), as well as those containing about 12 to 20% Cr and 25 to 67% Ni (such as HT, HU, and HX), usually are used in the as-cast condition. However, for applications that require maximum thermal fatigue properties, annealing is sometimes beneficial. Castings of alloy HT (35% Ni, 15% Cr), for example, may provide improved performance by being heated to 1040 °C (1900 °F), held at temperature for about 12 h, and furnace cooled, prior to being placed in service.

Solution Treating and Aging

The principal precipitation-strengthened iron-base alloys are A-286 and Incoloy 901. These two alloys, which contain titanium and aluminum, are strengthened mainly by precipitation of γ'. The γ' is metastable and transforms to η phase (Ni_3Ti) if the aluminum-titanium ratio is too low.

Iron-base alloys such as A-286 are relatively insensitive to the rate of cooling from the solution treating temperature (Table 11.5). Thus, section thickness is not of major concern in relation to cooling practice. The effect of varying amounts of cold work on grain growth in A-286 during solution treating is illustrated in Fig. 11.3. The initial material was in the solution treated condition with a maximum grain size of ASTM 5. Cold working 14% and re-solution treating reproduced this same grain size. However, cold working in the range of 1 to 5% caused excessive grain growth during subsequent solution treating.

Cold working affects the properties after aging, as illustrated in Fig. 11.4 for A-286. Diamond pyramid hardness after cold working and before aging, as well as the peak hardness after aging, increases greatly with

Table 11.5 Effect of cooling rate from 980 to 535 °C (1800 to 1000 °F) on short-time properties of A-286(a)

Cooling rate °C/min	Cooling rate °F/min	Yield strength(b) MPa	Yield strength(b) ksi	Tensile strength MPa	Tensile strength ksi	Elongation, %	Reduction in area, %
Tested at 20 °C (70 °F)							
Oil quench		680	99	1030	149	24	39
15	27	690	100	1030	150	22	42
7	15	680	99	1020	148	24	39
5	9	700	102	990	143	22	36
1	2	660	96	1030	149	23	33
Tested at 650 °C (1200 °F)							
Oil quench		650	94	780	113	8	10
5	27	650	94	770	112	6	7
7	15	630	92	770	112	9	16
5	9	590	86	740	108	7	17
1	2	220	90	740	108	11	18

(a) After cooling from the solution temperature, all samples were aged at 720 °C (1325 °F) for 16 h and air cooled. (b) At 0.2% offset

increasing amounts of cold rolling. The temperature at which peak hardness is attained decreases with increasing amounts of reduction. The temperature at which softening occurs also decreases with increasing amounts of deformation. As shown in Fig. 11.4, the material cold worked 81% and then aged at 760 °C (1400 °F) for 16 h is softer than material that was not cold worked before being aged.

To obtain uniformity of properties in parts that have been cold worked nonuniformly, higher-than-normal aging temperatures sometimes are used, followed by a second aging cycle at lower-than-normal aging temperature.

Practical Heat Treatment

Acceptable mechanical properties in precipitation-strengthened iron-base and nickel-base superalloys are not always accomplished as a result of the first solution treating and aging procedure tried for a given alloy or application. To develop specified properties, adjustment of the following heat treating variables frequently is required:

* Solution temperature or time
* Aging temperature
* Addition of an intermediate (stabilizing) aging treatment at a temperature higher than that of final aging
* Addition of a second (final) aging treatment at a lower temperature
* Adjustment of one or both aging temperatures in a double aging cycle
* Addition of a third aging treatment

For specific examples of these changes, see "Heat Treating of Heat-Resisting Alloys" in Volume 4 of the 9th Edition of *Metals Handbook*.

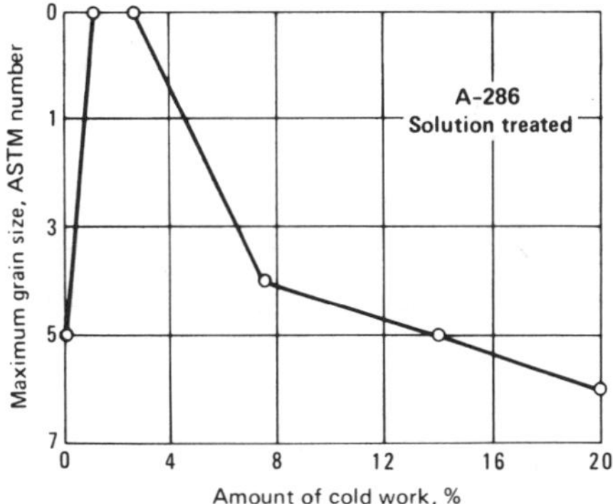

Influence of cold work on grain size of A-286 alloy solution treated at 900 °C (1650 °F) for 1 h and oil quenched.

Fig. 11.3 Effect of cold work

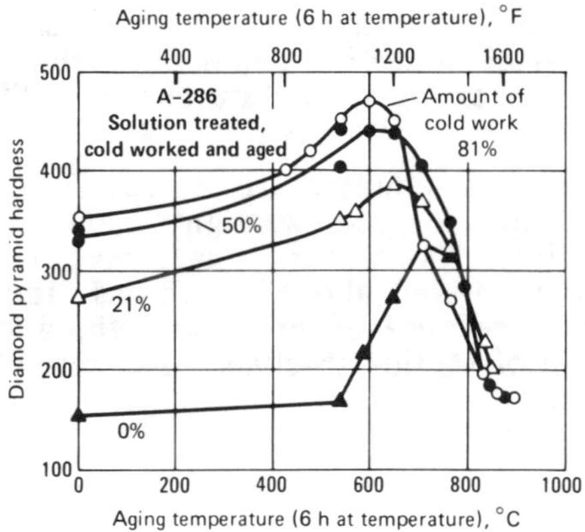

Fig. 11.4 Effect of cold work and aging on diamond pyramid hardness of A-286

HEAT TREATING OF NICKEL-BASE ALLOYS

Many nickel-base superalloys containing aluminum, titanium, and/or niobium are precipitation strengthened by formation of γ' or γ'' phase. Other nickel superalloys are of the solid-solution-strengthened type. Typical heat treating temperatures for select alloys are given in Tables 11.1, 11.2, and 11.3.

Stress Relieving and Annealing

These processes generally are applied to wrought solid-solution-strengthened nickel superalloys. Nickel-base alloy castings are seldom stress relieved or annealed. Such parts are used in the as-cast, solution treated, or solution treated and aged condition. The precipitation-strengthened

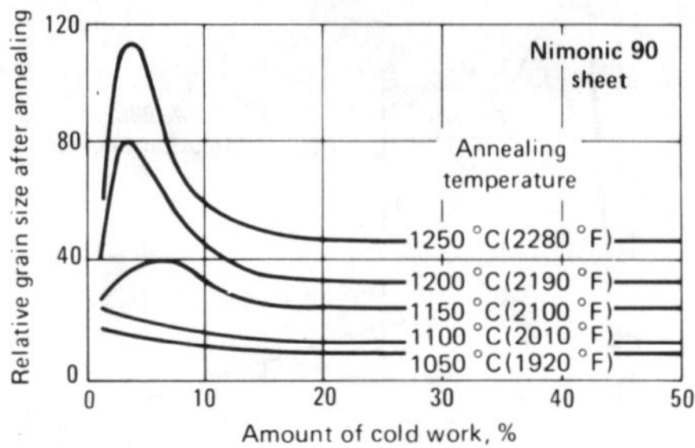

Nimonic 90 sheet was cold rolled in steps from 1.8 to 0.9 mm (0.072 to 0.036 in.) thick and annealed at five temperatures.

Fig. 11.5 Effect of cold work and annealing on grain size

wrought nickel-base alloys (such as Rene′ 41) are more crack-sensitive than the iron-base alloys and must be annealed during fabrication to relieve forming and welding stresses. As with other wrought alloys, grain size of nickel alloys after annealing depends on the amount of prior cold or hot work and the annealing temperature. This effect is illustrated for Nimonic 90 in Fig. 11.5.

Holding time at annealing temperature ranges from 1 to 2 h for most alloys. The more highly alloyed wrought materials such as Astroloy, Udimet 500, and Udimet 700 require longer holding times for dissolving microconstituents. For other alloys, such as Inconel 600, time at temperature is critical because of grain growth, and a holding time of about 15 min per inch of section thickness is commonly employed.

Room-temperature yield strength is sensitive to grain size, as illustrated in Table 11.6 for Rene′ 41 forged bar.

Atmospheres

Nickel-base alloys are susceptible to attack by sulfur-bearing atmospheres. Therefore, the gas used in a semimuffle furnace or for manufacturing exothermic atmosphere must have a very low sulfur content. Thin sheet is most likely to be damaged by sulfur-bearing atmospheres, or by oxidation in air. Argon inert environment atmospheres frequently are used to preclude excessive oxidation. Vacuum furnaces also may be used, provided the necessary cooling rates can be obtained.

Solution Treating

Solution treating temperatures of nickel-base precipitation-strengthened superalloys are determined primarily by the properties that are desired after subsequent aging. In wrought alloys, for optimum creep-rupture properties, higher solution treating temperatures are used than when optimum short-time tensile yield strength and elongation at elevated temperature are desired. For instance, Rene′ 41 and Udimet 500 are

Table 11.6 Relationship of grain size and yield strength of forged Rene′ bar

ASTM grain size	Average 0.2% yield strength MPa	ksi
3	840	122
5	900	130
7	1010	147
8	1060	154

solution treated at 1175 °C (2150 °F) for optimum creep-rupture properties and at 1080 °C (1975 °F) for optimum short-time tensile properties. For cast alloys, solution treatment temperatures primarily are a function of incipient melting temperatures, as well as furnace capability. Protective atmosphere or vacuum is necessary.

Because wrought nickel-base alloys are often solution annealed below the γ' solubility limit and because cast nickel-base alloys frequently are solutioned very close to the incipient melting temperature, control of furnace temperature is important when these materials are being heat treated. Variations in furnace temperature can result in failure to achieve specified properties. Should this occur due to inadequate solution heat treatment, a reheat treatment sometimes is effective in producing desired results. When incipient melting has occurred, reheat treatments are not able to re-establish adequate properties in the affected part.

Table 11.7 indicates how reheat treating provided turbine wheels of wrought Rene′ 41 with sufficient yield strength to exceed a minimum specification of 896 MPa (130 ksi), which was not consistently met by a single two-stage treatment under identical conditions. Re-treatment also increased both tensile strength and reduction in area. Elongation was not affected.

Higher solution heat treating temperatures and a faster quench provide more favorable properties for forming Rene′ 41 (Fig. 11.6). However, exposure to temperatures at which the carbides dissolve promotes subsequent formation of brittle films at the grain boundaries, with harmful effects on high-temperature properties.

Shrinkage During Aging

For design purposes, the contraction of components made of precipitation-strengthened nickel-base superalloys such as Rene′ 41 must be taken

Table 11.7 Effect of reheat treatment on mechanical properties of Rene′ 41 turbine wheels(a)

Treatment	Ultimate tensile strength MPa	ksi	Yield strength(b) MPa	ksi	Elongation in 50 mm (2 in.), %	Reduction in area, %
Heat treated(c)	1310-1380	190-200	850-920	123-133(d)	16-22	17-21
Re-heat treated(e)	1340-1420	195-206	920-1030	134-150(f)	16–22	18-24

(a) Data represent eight tests for each treatment; properties at 20 °C (68 °F). (b) Specified minimum. (c) Solution treated at 1065 °C (1950 °F), air cooled; aged at 760 °C (1400 °F) for 16 h, air cooled. (d) Average, 880 MPa (128 ksi). (e) Re-heat treated repeating procedure in (c). (f) Average, 980 MPa (142 ksi)

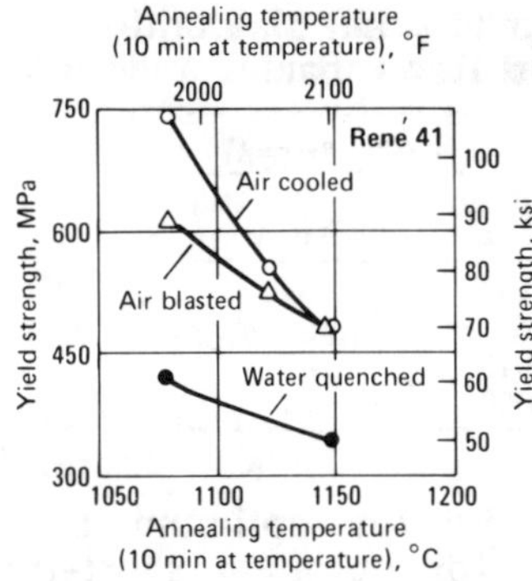

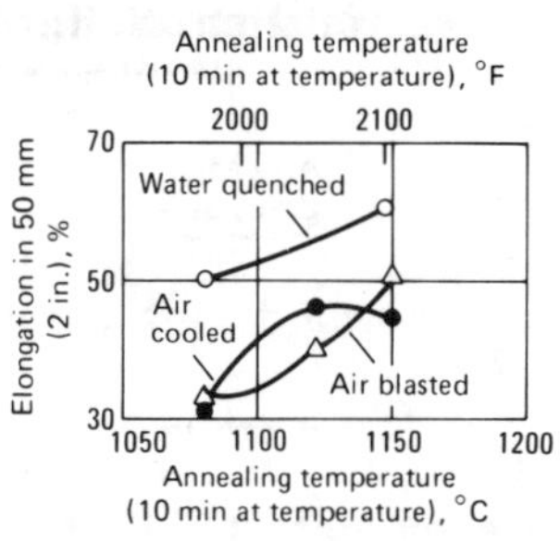

Fig 11.6 Effects of annealing temperature and quenching rate on typical room-temperature ductility and yield strength of Rene′ 41

into consideration. For example, an average contraction of 0.033 mm/mm (0.0013 in./in.) can be expected after aging Rene′ 41 for 16 h at 760 °C (1400 °F), or for 1 h at 900 °C (1650 °F) followed by 10 h at 760 °C (1400 °F). This average contraction rate reflects both longitudinal and transverse measurements on three heats of Rene′ 41, and the following treatments prior to aging:

* Solution treated at 1080 °C (1975 °F)
* Solution treated at 1080 °C (1975 °F), then annealed at 1105 °C (2025 °F), re-solutioned at 1080 °C (1975 °F)

Dilatometric measurements indicate that contraction is essentially complete within 6 h after the aging temperature is reached.

HEAT TREATING OF COBALT-BASE ALLOYS

Cobalt-chromium-nickel superalloys can be full annealed or stress relief annealed, but are rarely solution treated, because the γ'-type precipitates are not involved in strengthening of cobalt alloys. Aging heat treatments may be given, but the purpose of these would be to modify the carbide distribution. Heat treating temperatures for select alloys are given in Tables 11.1, 11.2, and 11.3.

Full annealing, rather than stress relieving, is recommended for most of the wrought cobalt-chromium-nickel alloys whenever high residual stresses are developed during fabrication. When fabrication is completed, however, these alloys can be safely stress relieved at about 55 °C (100 °F) below the annealing temperature (Table 11.1), provided excessive distortion does not occur.

Annealing of the cast cobalt-base alloys is often avoided, because of the undesirable precipitation of coarse carbides that occurs during slow cooling after annealing. Investment cast gas turbine buckets and guide vanes that are made to close dimensional tolerances and that must exhibit excellent resistance to stress cracking occasionally are stress relieved at 870 to 900 °C (1600 to 1650 °F) for 2 to 4 h. Cooling rate has no appreciable effect on the properties of annealed or stress relieved cobalt-base alloys.

Aging reactions of cobalt-base alloys such as HS-31 (X-40) are based on the random precipitation of various complex carbides and intermetallics. Cobalt superalloys are strengthened principally by the presence of carbide second phases, and consequently, the carbon content of these alloys is quite high. HS-31, a cast alloy containing 0.4% C, may contain $Cr_{23}C_6$, Cr_3,C_2, MC, or M_6C (or both), as well as μ (M_7W_6) phase. The ratio of chromium to carbon in the alloy determines which of the chromium carbides will appear during aging; a high ratio favors the precipitation of $Cr_{23}C_6$. Cobalt, iron, tungsten, and nickel can replace part of the chromium in $Cr_{23}C_6$. The $Cr_{23}C_6$ can be precipitated in different forms, represented by discrete, acicular, or lamellar shapes. Discrete particles are obtained by aging at temperatures near 760 °C (1400 °F). Aging at higher temperatures causes the appearance of the acicular precipitate. The third form, lamellar, resembles pearlite and is nucleated by cooling from a solution treating temperature between 1205 °C (2200 °F) and the temperature range where $Cr_{23}C_6$ precipitates.

Cobalt-base alloys generally are not susceptible to attack by furnace atmospheres and air environments can be used for heat treatment. At higher temperatures or in the case of vacuum melted alloys that contain reactive elements such as zirconium (for example, MAR-M 509), a vacuum, inert, or almost neutral environment may be required.

Cobalt-base alloys are welded, either as fabrications, or to repair weld cracks in base alloys after service operation. Stress relief heat treatments are applied after welding, but there is no uniform preweld heat treating practice.

Chapter 12

Cleaning, Finishing

MACHINING

Machining is an important process in the manufacture of superalloy parts. Despite advances in near-net shape processing via precision casting, precision forging, and powder metallurgy (P/M) processing, machining plays a vital, albeit expensive, role in superalloy part manufacture. Much of the high machining cost is due to the fact that cutting speeds for superalloys are $^1/_{10}$th to $^1/_{20}$th of those used for steel.

Single-point turning is the most frequently used machining process for superalloys, followed by grinding, end milling, and broaching, because the conventional chip-making processes provide much higher metal-removal rates than processes such as electrochemical machining (ECM). Other chip-making processes used for superalloys are planing, drilling, screw machining, boring, gear cutting, and tapping.

Machining Properties

Superalloys generally are classified as having poor machinability. This is not surprising, because the very characteristics that render superalloys good high-temperature materials are responsible for their poor machining behavior. In general, superalloys possess the following mechanical/machining properties:

* Strength is retained at high temperatures, whereas common tool steels begin to soften.
* They possess unusually high dynamic shear strength.
* The presence of hard carbides makes superalloys abrasive.
* Work hardening occurs during metal cutting.
* They possess poor thermal diffusivity, thus leading to high temperatures at the tool tip.
* A tough, continuous chip is produced during metal cutting.

Most of the conventional means of improving machinability are not effective with superalloys. Alloy modification and heat treatment generally are not effective. Hot machining holds some possibilities, but it is expensive and introduces other problems. Use of nonconventional electrically assisted processes is another approach, but metal-removal rates are very low. Obviously, the solution to this general problem must come from the development of new cutting tool materials.

CUTTING TOOLS

Commonly used cutting tool materials are high speed steels, cast alloys, carbides, and ceramics. For superalloys, high speed cobalt tool steels are recommended for milling, drilling, tapping, and broaching. Carbides are used for turning, planing, and face milling. The most commonly used carbide is the C-2 grade (91% WC and 9% Co). Modification of tungsten carbide tools with the addition of 0.5 to 4% tantalum carbide has been beneficial in improving abrasion resistance. Titanium carbide tools are not applicable for superalloys, because of the high solubility of titanium carbide in nickel and cobalt.

Ceramic tools with their lack of deformability are not suitable for machining superalloys, because they exhibit edge chipping and premature failure. The common high speed steels and sintered carbides used for superalloy machining are listed in Tables 12.1 and 12.2, respectively.

Tool Life

In machining superalloys, the common causes of tool failure are excessive flank wear, excessive groove formation at chip edges, and the inability to meet surface finish and accuracy requirements. Other contributing factors include excessive crater depth and destruction of the cutting edge by fracture, plastic flow, or melting. Surface speed is the single most important factor in determining tool life. In contrast with other workpiece materials, feed and depth of cut are also important for tool life in machining superalloys. Cutting temperatures for superalloys are typically 760 to 1010 °C (1400 to 1850 °F). These temperatures are sufficiently high that oxidation and diffusion become significant contributing factors to total tool wear even for the carbide tools.

Considerable data are available on typical carbide tool lives for turning various superalloys relative to cutting speeds. For example, A-286, an iron-base superalloy, is the easiest of the superalloys to machine, with a tool life of 25 min at a cutting speed of 0.80 m/s (155 ft/min). The poorest machining characteristics are exhibited by IN-100, a cast nickel-base airfoil superalloy. This alloy has a typical tool life of only 12 min at the reduced cutting speed of 0.15 m/s (30 ft/min). Falling between these two extremes in machinability characteristics, the following superalloys are ranked in order of increasing ease of machining: MAR-M 302 (cobalt base), AF2, Rene′ 95, U-500, Rene′ 41 (all nickel base), HS-25 (cobalt base), INCO 718, Waspaloy, and Incoloy 901 (all nickel base).

Similar data are available for carbide face milling, high speed drilling, and other machining operations. The current literature also provides recommendations for all machining operations for various superalloys, as well as tool geometries. These data, however, are only a general guide. Machining of superalloys is so difficult that careful study should be undertaken to develop a set of machining parameters that results in reasonable tool life, along with an economic analysis covering speeds, feeds, tool materials, and cutting tool reconditioning costs.

Table 12.1 High speed tool steels for machining superalloys

Type	Composition, %						Applications
	C	W	Mo	Cr	V	Co	
M6	0.8	4	5	4	1.5	12	Heavy cuts, abrasion resistant
M30	0.8	2	8	4	1.25	5	Heavy cuts, abrasion resistant
M33	0.9	1.5	9.6	4	1.15	8	Heavy cuts, abrasion resistant
M34	0.9	2	8	4	2	8	Heavy cuts, abrasion resistant
M36	0.8	6	5	4	2	8	Heavy cuts, abrasion resistant
M41	1.1	6.75	3.75	4.25	2	5	Heavy cuts, abrasion resistant
M42	1.1	1.5	9.5	3.75	1.15	8	Heavy cuts, abrasion resistant
M43	1.2	2.75	8	3.75	1.6	8.25	Heavy cuts, abrasion resistant
M44	1.15	5.25	6.25	4.25	2.25	12	Heavy cuts, abrasion resistant
M46	1.25	2	8.25	4	3.2	8.25	Heavy cuts, abrasion resistant
M47	1.1	1.5	9.5	3.75	1.25	5	Heavy cuts, abrasion resistant
T4	0.75	18	...	4	1	5	Heavy cuts
T5	0.8	18	...	4	2	8	Heavy cuts, abrasion resistant
T6	0.8	20	...	4.5	1.5	12	Heavy cuts, hard material
T8	0.75	14	...	4	2	5	General purpose, hard material
T15	1.5	12	...	4	5	5	Extreme abrasion resistant

Table 12.2 Sintered carbides for machining superalloys

Grade	Composition, %			Transverse rupture strength	
	WC	TiC	Co	MPa	ksi
C-1	94	...	6	2186	317
C-2	91	...	9	1896	275
C-3	95.5	...	4.5	1379	200
C-4	97	...	3	1207	175
C-6	82	8	10	1482	215
C-7	80	12	8	1207	175
C-8	84	10	6	1427	207
C-50(a)	72	8	8.5	...	...

(a) Also contains 11.5% TaC.

GRINDING

Grinding is characterized by very high surface speeds, high forces, and very small chip size, approximately 1% of the thickness involved in fine turning. The energy per unit volume of metal removed is high, about 30 times that for turning. Consequently, the cutting material must be very hard, refractory, and nonreactive. It is very important that the grinding wheel cuts rather than deforms. Increasingly with superalloys, grinding is a shaping technique as well as a finishing operation. Primarily Al_2O_3 abrasive, usually vitreous rather than resinoid bonded, is used in grinding superalloys. Silicon carbide, diamond, and Borazon, however, have advantages for some applications.

Although grinding is used primarily to produce a finished component with good surface integrity, productivity and economy also are important factors. Wheel wear is measured by a grinding (G) ratio, i.e., the volume of metal removed from the workpiece to the volume of the grinding wheel that is worn away during the process. In grinding superalloys, the G ratio varies from 2 to 70, whereas in grinding steel, the G ratio is between 60 to 200. For example, the approximate G ratios for IN-100, HS-25, Waspaloy, INCO 718, and Rene′ 41 are 4, 6, 10, 11, and 18, respectively.

Finish grinding parameters selected for good surface integrity tend to promote high wheel wear and low G ratios. For rough grinding, however, conditions can be selected to give higher G ratios. Rough grinding conditions still should not produce grinding cracks. Wrought superalloys usually can be handled in a manner to prevent this phenomenon, but the high-strength, cast nickel-base and cobalt-base alloys are crack sensitive. Special precautions, therefore, must be taken in grinding these alloys.

CLEANING

Cleaning is required to remove contaminants from the surfaces of parts made of superalloys. Shop soils such as oil, grease, and cutting fluids can be removed by conventional solvents or soaps. Metallic contaminants, tarnish, and scale resulting from hot working or heat treating operations must also be removed.

REMOVAL OF METALLIC CONTAMINANTS

Superalloy parts may accumulate traces of other metals on their surfaces as a result of contact with cutting tools, forming dies, and heat treating fixtures. In some instances, these metals are not harmful, but in others, the presence of contaminants can have a highly deleterious effect. The high-temperature strength and ductility of nickel-base γ'-strengthened superalloys are severely reduced by minute quantities of lead, bismuth, antimony, selenium, and arsenic. Small amounts of aluminum, even though it is a γ' element, readily alloy with nickel-base superalloys at elevated temperatures and degrade the affected area. Also, copper and silver reduce the high-temperature strength of nickel superalloys.

In operations such as cutting and forming, metal contamination can be avoided or sharply reduced by the use of lubricants. This is preferred practice, because lubricants can be removed easily and cheaply. If contamination cannot be avoided, testing is recommended to determine the seriousness of the contamination. After heat treating operations, metallographic tests to detect alloy diffusion or bend tests to detect embrittlement commonly are used. Lead may be detected by the yellow precipitate that forms when the following test solution at 21 to 60 °C (70 to 140 °F) is applied to the suspected surface:

Chromic acid..........................10 wt%
Sodium chlorate......................1.5 wt%
Water..88.5 wt%

Although metal contamination is not always harmful, all surface contaminants should be removed before superalloys are heat treated or subjected to service at elevated temperatures.

Mechanical Removal

Dry or wet abrasive blasting with metal-free abrasives is an effective means of removing metal contamination, as are polishing with ceramic materials and wet tumbling. The shape of a part, the surface finish required, and the allowable loss of gage will determine the suitability of these mechanical methods.

Chemical Removal

Chemical methods are used more often than mechanical methods for removing metal contamination. A typical procedure for the chemical removal of iron, zinc, and thin films of lead is as follows:

* Vapor degrease or alkaline clean
* Immerse part in a 1-to-1 solution (by volume) of nitric acid (1.41 sp gr) and water for 15 to 30 min at about 35 °C (95 °F)
* Water rinse and dry

Another procedure, which has proved successful for removing brass, lead, zinc, bismuth, and tin from nickel-base and cobalt-base alloys, is given below:

* Vapor degrease or alkaline clean
* Soak at room temperature in solution of:

  Nitric acid........................54 g/L (7.22 oz/gal)
  Acetic acid.......................150 to 375 g/L (20 to 50 oz/gal)
  Hydrogen peroxide..........19 to 64 g/L (2.5 to 8.5 oz/gal)

Soaking time will vary from 20 min to 4 h, depending on the severity of the contamination, and is determined by visual observation of the reaction. After this treatment, parts must be rinsed thoroughly in water and dried. When possible, a test specimen should be immersed for the maximum time anticipated and examined for chemical attack before processing the first load of parts.

Nickel-base superalloys should be acid etched to prepare for subsequent nondestructive inspection. The etching process removes smeared metal that may be present as a thin surface layer following machining and/or shot blast cleaning. Parts to be examined may be etched by immersion in a bath of 80% hydrochloric acid, 13% hydrofluoric acid, and 7% nitric acid to remove the disturbed or smeared layer. This bath may leave smut that must be removed in a second bath containing 22% iron chloride, 75% hydrochloric acid, and 2% nitric acid and water. Following rinsing and drying, the parts are ready for optical macro-inspection and penetrant inspection. The extent of etching depends on the depth of the smeared layer. All of the disturbed layer should be removed, but over-etching causes excessive penetrant retention.

REMOVAL OF TARNISH

Tarnish, or thin oxide film, may not always have a harmful effect on the end use of parts made of superalloys. It may even prove useful. For example, tarnish may function as a bond for paint, as a barrier to prevent diffusion from another alloy, or as a retardant to further oxidation. However, considerations such as functional requirements may indicate that tarnish should be removed from parts.

Tarnish always should be removed before welding or brazing. Abrasive cleaning methods such as those used for removing metallic contaminants also are used for removing tarnish. The applicability of these methods is determined by the shape of the parts, the surface finish required, and the allowable loss of gage.

Flash pickling is used more often than abrasive cleaning for tarnish removal, because abrasive cleaning removes some metal and can degrade the surface finish. A typical formula for flash pickling is:

Nitric acid (1.41 sp gr).....................23 vol%
Hydrofluoric acid (1.26 sp gr)..........4 vol%
Water..73 vol%

Parts are immersed in the above solution at about 52 °C (125 °F) for 1 to 5 min. Warming the parts in hot water before flash pickling speeds tarnish removal. Water rinsing and drying must follow flash pickling.

REMOVAL OF REDUCED OXIDE AND SCALE

Acid pickling, abrasive cleaning by tumbling or blasting, and descaling in molten salt are the most widely used methods for removing reduced oxide or scale from superalloys, listed above in order of decreasing preference based on economic considerations. Alkaline scale conditioning is helpful in modifying scale so that it may be more readily removed by these methods. When extremely heavy oxide layers must be removed, grinding is an appropriate preliminary operation. Combinations of two or more methods are often used.

Reduced Oxide Films

Reduced oxide films are formed on parts that are heated in reducing atmospheres out of contact with air. Sometimes, these oxides can be removed by immersing parts for 5 to 15 min in a tarnish-removing flash pickling bath. (See above formulation for flash pickling.) However, most superalloys, because of their high content of oxide-forming metals such as nickel, cobalt, and chromium, form a tenacious coating in the presence of carbon monoxide or water. The oxides formed vary widely with composition of the alloy and furnace atmosphere. Usually, pickling is required for their removal. Scale conditioners facilitate removal of these oxides. (See typical procedure in Table 12.3.)

Scale

Scale develops on hot forged, hot formed, or heat treated parts that are processed in air. Usually, scale is tenacious and occurs in all gradations, including severe coatings that result from heating in an oxidizing furnace using high-sulfur fuels. Scale formed under such conditions has a dull, spongy appearance. Fine cracks may be present, and patches of scale may break from the surface. The roughened underlying metal cannot be smoothed by pickling. In these extreme conditions, grinding or abrasive blasting to sound metal, followed by flash pickling, is recommended.

Table 12.3 Procedure for removing scale from superalloys

Operation	Time	Temperature °C	Temperature °F	Solution	Concentration
Precleaning cycle					
Vapor degreasing	5-10 min	87-88	185-190	Stabilized trichlorethylene	...
Emulsion cleaning	10-20 min	54-66	130-150	Emulsion cleaner	20 vol %
Scale-conditioning cycle					
Alkaline chelating	15-30 min	125-135	260-275	Caustic solution containing alkanol amines and aliphatic hydroxy acids(a)	...
Alkaline oxidizing	1-2 h	95-105	205-220	Potassium permanganate	5 wt %
				Sodium hydroxide	20 wt %
				Water	rem
Pickling cycle					
Acid pickling	5-30 min	49-60	120-140	Hydrofluoric acid	4 wt %
				Nitric acid	18 wt %
				Water	rem

(a) For use and composition refer to U. S. Patents 2,843,509, 2,992,946, 2,992,995, and 2,992,997

Scale Conditioning

Scale conditioning is used to soften, modify, or reduce scale for easier and more uniform acid pickling, but is seldom required for removal of discoloration or interference coatings. A scale conditioning bath consists of a highly alkaline aqueous solution, sometimes containing complexing and chelating compounds. The main purpose of these agents is to solubilize the scale as much as possible. The performance of a particular chelating agent depends on the affinity of the compound for the metal ions present, the pH of the scale conditioning solution, and the physical and chemical composition of the scale.

Minimal scale removal occurs during treatment in the alkaline scale conditioning bath. Further treatment in highly alkaline solutions containing a strong oxidizing material, such as potassium permanganate, is often necessary. Scale on heat-resistant alloys sometimes contains carbon and incompletely burned and polymerized residues in addition to metallic oxides. These organic components react with the oxygen released by the alkaline oxidizing bath.

Acid Pickling

The conditioned scale finally is subjected to acid pickling, during which most of the high-temperature scale breaks away or becomes so loosely attached that pressure rinsing with water completes descaling. The acid pickle is usually a dilute nitric acid or a hydrofluoric acid–nitric acid solution. In addition to removing scale, pickling solutions containing nitric acid remove many surface contaminants through oxidation. However, because the acid solution attacks the basis metal, it is necessary to limit the time of pickling to prevent excessive metal loss and excessive roughening of the metal surface.

A special procedure is used for treating γ'-strengthened nickel-base superalloys that are high in aluminum and titanium before welding or brazing. When the alloy is in the solution treated condition and descaling is required, it is treated in a scale conditioning solution, as described above, after which it is immersed in a solution of 30 wt% nitric acid and 3 wt% hydrofluoric acid for 5 to 10 min. Alloys in the aged condition are descaled anodically in an acid solution. A solution for this procedure contains 75 wt% sulfuric acid and 3 wt% hydrofluoric acid and is operated with a current density of 2 to 4 A/dm^2 (20 to 40 A/ft^2) and graphite cathodes. The material is immersed in the electrolytic cleaning bath for 3 to 12 min; the operation is complete when the amperage drops almost to zero. Sodium sulfite (1.6 wt%) is used to reactivate the solution after a period of operation.

Superalloys containing less than about 12% Cr can undergo high metal loss or develop intergranular attack in pickling. When the susceptibility of a material to excessive metal loss or intergranular attack by acids is unknown, mechanical descaling is safer than acid pickling. In one instance, it was proved that acid descaling caused intergranular attack and subsequent loss of ductility in aged Rene′ 41, which contains 19% Cr.

Weld areas normally vary in composition and structure from the basis metal and do not react to the conditioning and pickling cycles in the same manner as the basis metal. Weld areas or the heat-affected zones often are susceptible to selective attack during pickling. Although inhibitors may eliminate or reduce selective attack, abrasive or other mechanical descaling methods are preferable to acid pickling for removing scale from welded parts, unless a safe pickling procedure has been found for a given application.

Hydrogen embrittlement does not occur in superalloys as a result of aqueous descaling. Details of typical multicycle descaling operations are given in Tables 12.3 and 12.4.

Abrasive Blasting

Abrasive blasting with dry aluminum oxide can be used to remove oxide and scale from all types of wrought and cast superalloys. Silicon carbide is more expensive than aluminum oxide and is seldom used. Silica (silicon dioxide or sand) has limited application because of its lack of cutting ability.

Metallic shot and grit should not be used for descaling superalloys, unless their use is followed by pickling to remove metal contamination. For parts to be welded or brazed, or where furnace atmospheres have produced highly tenacious scale, pickling after dry abrasive cleaning is recommended, regardless of the abrasive used.

Wet abrasive blasting is also used for cleaning heat-resistant alloys. This process uses silica abrasive particles (No. 200 to No. 1250) that are 0.074 to 0.010 mm (0.0029 to 0.0004 in.) in diameter mixed with water to produce a slurry that removes loose scale, discoloration, and soils. Metal loss is not excessive when normal pressures and exposure times are used.

Surfaces that have been wet blasted are usually suitable for welding, brazing, electroplating, and final inspection; further cleaning is seldom necessary. Exceptions are the γ'-strengthened nickel-base superalloys with high combined titanium and aluminum contents, which require the special procedure previously discussed.

Most of the general advantages and limitations of abrasive cleaning of steel also apply to superalloys. An exception is the risk of contamination from metallic abrasives or from cleaning parts with abrasives that have been used for other metals of widely different compositions. For example, superalloys should not be blasted with abrasive material that has been used for cleaning low-alloy steel, aluminum, copper, or magnesium. However, abrasives that have been used to clean titanium and corrosion-resistant steels have been used for cleaning superalloys without serious contamination. Flash pickling of superalloys after abrasive cleaning provides additional assurance that no harmful surface contamination remains.

Wet Tumbling

Wet tumbling by the barrel or vibratory method can be used for descaling superalloys, if the parts are of suitable shape and size. Removal of burrs and sharp edges is accomplished in the same operation. Shop soils are also removed, thus eliminating the need for preliminary degreasing.

Parts are tumbled or vibrated in a mixture of acid descaling compound and metal-free abrasives, after which they undergo a neutralizing cycle.

Table 12.4 Procedure for removing scale from Inconel alloys

Operation	Time	Temperature °C	Temperature °F	Solution(a)	Concentration(a)
Alkaline conditioning	1-2 h	96-105	205-220	Sodium hydroxide	20 wt %
				Potassium permanganate	5 wt %
Rinse	15-30 s	Not heated	Not heated	Water quench and water spray	...
Acid pickling	5-10 min	60-71	140-160	Sulfuric acid (1.83 sp gr)	7.5 vol %
				Hydrochloric acid (1.16 sp gr)	12 vol %
Rinse	15-30 s	Not heated	Not heated	Water	...
Acid pickling	10-20 min	60-71	140-160	Nitric acid (1.41 sp gr)	20 vol %
Rinse	15-30 s	Not heated	Not heated	Water	...
Acid pickling	5-60 min(b)	49-54	120-130	Hydrofluoric acid (1.26 sp gr)	3.7 vol %
				Nitric acid (1.41 sp gr)	22 vol %
Rinse	15-30 s	Not heated	Not heated	Water	...

(a) Remainder is water. (b) Type of oxide will determine immersion time required; until immersion is established, inspect frequently to avoid overpickling

Table 12.5 Procedure for sodium hydride descaling and acid pickling of superalloys

Operation	Time, min	Temperature °C	Temperature °F	Solution(a)	Concentration(a), %
Sodium hydride descale	5-20	370-390	700-730	Sodium hydride	1.5-2.0
Quench	¼-½	Not heated	Not heated	Water	...
Neutralizing rinse	1-3	Room temperature to 60	Room temperature to 140	Sulfuric acid	2-10
Brightening pickle	5-15	54-60	130-140	Hydrofluoric acid	2-4
				Nitric acid	15-20
Rinse	¼-½	Not heated	Not heated	Water	...
High-pressure spray wash	(a)	Not heated	Not heated	Water(b)	...

(a) Governed by shape of part; 1 min on parts with accessible surfaces. (b) Water pressure of 690 kPa (100 psi)

Precautions regarding metal contamination are similar to those noted above for abrasive blasting.

Pickling is required after tumbling and before resistance welding to remove residual smut, which can cause poor-quality weldments. There is less need for pickling prior to fusion welding, unless inspection of weldments reveals porosity or inclusions caused by the tumbling process.

Wire Brushing

Wire brushing sometimes is used for removal of very light scale or surface discoloration. Brushes, when used for superalloys, must have stainless steel bristles.

Salt Bath Descaling

Salt bath descaling is an effective first step in removing scale from superalloys. This process is generally more costly than acid pickling, particularly if production is intermittent, because of the cost of maintaining the bath during idle time. The electrolytic salt bath used for descaling superalloys does not contain sodium hydride, but uses fused caustic soda. The workpiece and tank are alternately negative and positive poles of a direct current.

A fused caustic soda bath containing oxidizing salts such as sodium nitrate has been used as the initial step in descaling superalloys. This oxidizing bath, which is operated at 425 to 540 °C (800 to 1000 °F), is slightly more effective than the sodium hydride bath for high-chromium alloys such as the complex cobalt-chromium-nickel alloys. Processing steps are similar. The work load is immersed in the oxidizing bath for 5 to 15 min, quenched in water, soaked in a solution of 5 to 10% sulfuric acid at 71 °C (160 °F) for 1 to 5 min, and then dipped in a solution of 15 to 20% nitric acid and 2 to 4% hydrofluoric acid at 54 to 60 °C (130 to 140 °F) for 2 to 15 min. A typical sequence for hydride descaling and pickling is given in Table 12.5.

FINISHING PROCESSES

Many finishing operations commonly used for steel and other metals are not required for superalloys. This is due to: (*a*) the inherent corrosion resistance of superalloys in a wide range of environments, and (*b*) the end use of superalloy parts frequently does not require a polished finish.

The oxide coating obtained during processing is frequently of value to superalloys that are subjected to elevated temperatures in service. Consequently, the dense, tenacious oxide that forms on formed or machined finished parts during final heat treatment is allowed to remain as protection against further oxidation.

Typical finishing processes include:

- Polishing
- Electroplating
- Ceramic coating

* Diffusion coating
* Shot peening

For more details on these processes, see "Cleaning and Finishing of Heat-Resistant Alloys" in Volume 5 of the 9th Edition of *Metals Handbook*.

Chapter 13

Welding

INTRODUCTION

Superalloys, except those with high aluminum and titanium contents, are welded with little difficulty. Nickel-base superalloys that have a slow aging reaction, for example, INCO 718, also are welded without problems. Most welding concern is focused on the high-strength γ'-strengthened nickel superalloys and high precipitation-strengthening-element (aluminum + titanium) content alloys. The procedures used in welding superalloys depend to some extent on the mechanism by which they are strengthened for high-temperature service - primarily solid-solution strengthening or precipitation strengthening.

Many superalloys are susceptible to cracking during welding or during subsequent heat treatment. The severe service conditions to which superalloys are exposed make changes in microstructure and properties, as produced by arc welding, of special importance and make avoidance of cracking imperative. The strength of fusion welds in superalloys usually is lower than the strength of the base metal, and this has a deleterious effect on long-term weld performance.

Selection of an appropriate filler metal to be used when welding a specific superalloy is critical for success. As a general rule, the manufacturers' recommendations should be followed. Furthermore, the ability to weld superalloys with matching filler metals has been hampered by the lack of suitable compositions in fine-diameter wire, as well as by the inherently limited usefulness of superalloy compositions for use as weld metals. Nominal compositions of commonly arc welded superalloys are given in Table 2.1, Chapter 2.

WELDING PROCESSES

Superalloys can be welded by all of the arc welding processes. Gas tungsten arc welding (GTAW) is widely used, especially for joining thin sections. In general, shielded metal arc welding (SMAW) and gas metal arc welding (GMAW) are used for joining sections that are more than 6.4 mm (1/4 in.) thick, where the heat input does not adversely affect the weld metal or the base metal. Submerged arc welding (SAW) generally is used in high-volume production welding of sections that are more than 25 mm (1 in.) thick.

Table 13.1 Conditions for GTAW of superalloys(a)

Base-metal thickness, in.	Diameter of filler metal, in.(b)	Electrode diameter, in.(c)	Shielding gas — Gas	Shielding gas — Flow rate, ft³/h	Welding current, A(d)
0.010	0.020	0.040-0.060	Ar	12-15	10-15
0.020	0.030	0.060	Ar	12-15	15-25
0.030	0.030; 0.045	0.060	Ar	12-15	25-35
0.045	0.045	0.060	Ar	12-15	40-50
0.050	0.045	0.060	Ar	12-15	45-55
0.060	0.045	0.060	Ar	12-15	55-65
0.080	0.060	0.060	Ar	12-15	75-85
0.100	0.060; 0.090	0.093	Ar or He	12-20	95-105
0.125	0.060; 0.090	0.093	Ar or He	12-20	110-135
0.250	0.060; 0.090	0.093	Ar or He	12-20	130-200

(a) The data in this table are intended to serve as starting points for the establishment of optimum machine settings for welding workpieces on which previous experience is lacking. The data are subject to adjustment as necessary to meet the special requirements of individual applications. Torch nozzle diameter was 7/16 in.; nozzle had a gas lens. (b) Minimum wire diameters were applicable. (c) EWTh-2 electrodes. (d) DCEN with high-frequency arc starting. An increase of 10 to 20 A may be needed for melt-through T-joints.

The data in Table 13.1 are intended to serve as starting points for the establishment of machine settings for GTAW. These conditions are, in general, suitable for welding nickel, iron-nickel-chromium, iron-chromium-nickel, and cobalt-base superalloys, when making butt, corner, or T-joints, using an appropriate groove design based on stock thickness and application. An increase in welding current of 10 to 20 A may be needed for melt-through T-joints. Generally, the interpass temperature should range from 93 to 175 °C (200 to 350 °F), depending on the alloy. Oscillation of the welding torch may help to prevent cracking by changing the solidification pattern. This may also improve the appearance of the weld.

Cobalt Alloys

Cobalt-base superalloys are readily welded by gas metal arc (GMA), or gas tungsten arc (GTA) techniques. Cast alloys such as WI-52 and wrought alloys such as Haynes 188 are welded extensively. Filler metals generally are less highly alloyed cobalt-base alloy wire, although parent rod or wire (the same composition as the alloy being welded) has been used. Cobalt-base superalloy sheet also can be welded successfully by resistance techniques. Gas turbine vanes that crack in service are repair welded using the above techniques - for example, WI-52 vanes using Haynes 25 filler rod and 540 °C (1000 °F) preheat. Indeed, good weldability is an important reason for the selection of cobalt superalloys for this application. Appropriate preheat techniques are needed in GMA and GTA welding to eliminate tendencies to hot cracking. Electron beam welding (EBW) and plasma arc welding (PAW) can be used on cobalt-base superalloys, but usually are not required in most applications because alloys of this class are so readily weldable.

Nickel and Iron-Nickel Alloys

Precipitation-strengthened nickel and iron-nickel superalloys are considerably less weldable than cobalt-base superalloys. Because of the presence of the strengthening phase, these alloys tend to be susceptible to hot cracking (weld cracking) and postweld heat treatment (PWHT) cracking.

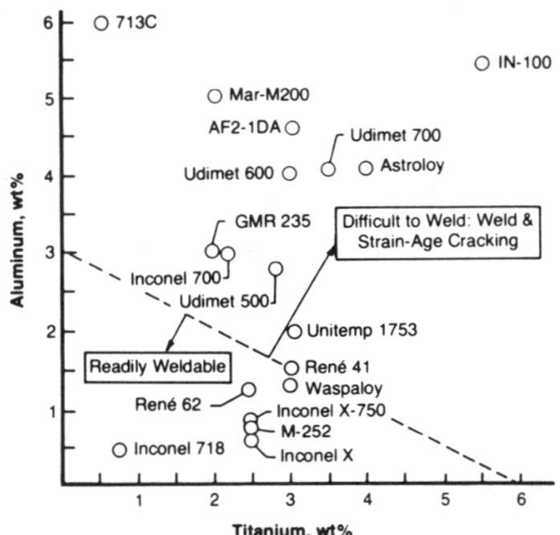

Fig. 13.1 Weldability diagram for some γ'-strengthened iron-nickel and nickel-base superalloys, showing influence of total aluminum + titanium hardeners

The susceptibility to hot cracking is directly related to the aluminum and titanium contents (γ' formers), as shown in Fig. 13.1. Hot cracking occurs in the weld heat-affected zone (HAZ), and the extent of cracking varies with alloy composition and weldment restraint.

Nickel and iron-nickel superalloys have been welded by GMAW, GTAW, EBW, laser, and PAW techniques. Filler metals, when used, usually are weaker, more ductile austenitic alloys so as to minimize hot cracking. Occasionally, base-metal compositions are employed as fillers. Such welding is generally restricted to the low volume fraction γ' alloys, usually in the wrought condition. Cast alloys of high volume fraction γ' have not been welded successfully on a consistent basis when filler metal is required, as in weld repair of service parts. However, electron beam welding can be used to make structural joints in such alloys. Friction or inertia welding also has been successfully applied to the lower volume fraction γ' alloys.

Because of their γ'-strengthening mechanisms and capabilities, many nickel and iron-nickel superalloys are welded in the solution heat treated condition. Special preweld heat treatments have been used for some alloys. Inconel 718 has unique welding characteristics in that its hardening phase (γ'') is precipitated more sluggishly at a lower temperature than is γ' so that the attendant welding-associated strains that must be redistributed are more readily accommodated in the weld metal and heat-affected zone. This alloy is welded in the solution treated condition and then given a postweld stress-relief and aging treatment that causes γ'' precipitation. Some alloys, such as A-286, are inherently difficult to weld despite only moderate levels of γ' hardeners. There is some evidence that high-titanium alloys may be more difficult to weld than alloys of similar γ' volume that rely on high aluminum-titanium ratios for their strength capabilities.

Weld techniques for superalloys must address not only hot weld cracking, but PWHT cracking, as well. This is of particular interest in terms of the microfissuring (microcracking) phenomenon that sometimes occurs. Microfissures can occur in the subsurface of the weld and therefore are difficult to detect. Tensile and stress-rupture strengths may be affected minimally by microfissuring, but fatigue strength can be drastically reduced.

In addition to being weldable by the usual fusion welding techniques discussed above, nickel and iron-nickel alloys can be resistance welded when in sheet form. Diffusion bonding, which relies on the application of pressure at the joint during the heating cycle to produce high-quality welds that match base-metal properties, has been used infrequently with superalloys.

CLEANING OF WORKPIECES

The weldability of superalloys is markedly affected by the cleanliness of the base metal and the filler metal. Shop dirt, paint, grease, oil, machine lubricants, processing chemicals, temperature-indicating sticks, marking crayons, oxide films, and scale are the main surface contaminants. Sulfur and lead in foreign material on the workpiece surface can diffuse into the base metal when it is heated and result in severe cracking.

Grease and oil are removed by commercial solvents and by vapor degreasing. Soaps can be removed by rinsing in hot water. Soluble oils, tallow fats, and fatty acids require a more complex cleaning procedure.

For more information on the removal of surface dirt, tarnish, oxide films, and heavy scale, see "Cleaning and Finishing of Heat-Resistant Alloys" and "Cleaning and Finishing of Nickel and Nickel Alloys" in Volume 5 of the 9th Edition of *Metals Handbook*.

Dust that has settled on the workpieces after cleaning can be removed by carefully wiping the joint area with clean, lint-free cloths dampened with a solvent such as methyl ethyl ketone. To avoid surface contamination of cleaned workpieces, white gloves are sometimes used during handling. After welding, all slag must be removed. If any surface slag left on a weldment should become molten during service, severe attack on the metal may occur.

WELDING FIXTURES

Fixtures used in arc welding superalloys are generally similar to those used on other metals. Chill bars are often used to cool the weld area rapidly. Backing bars, inserts, and facing plates that are in contact with the workpieces on either the root or the face side of the welds should be located so as not to contaminate the weld metal or base metal, or cause gas or flux entrapment. These accessories are usually made of copper.

Hold-down bars and backing bars extend the full length of the weld. The backing bars usually contain passages to facilitate inert gas shielding. When grooved backing bars are used, the grooves should be shallow to minimize melt-through and to limit the height of the root reinforcement. Grooves in backing bars should have rounded corners, causing them to be elliptical in shape, to prevent entrapment of slag.

NICKEL SUPERALLOYS

Some nickel-base superalloys are solid-solution strengthened, and others are strengthened by the precipitation of a second phase from a supersaturated solid solution. Many of the solid-solution nickel-base alloys, including Inconel 625 and Hastelloy C-276, are widely used in applications requiring corrosion resistance at temperatures ranging from cryogenic to 425 °C (800 °F) or higher; these alloys, therefore, are classified as being both heat resistant and corrosion resistant.

Powder metallurgy (P/M) nickel-base alloys may be strengthened by the uniform dispersion of refractory oxide particles throughout the matrix. This process is referred to as oxide dispersion strengthening (ODS). When these P/M materials are fusion welded, the oxide particles will agglomerate during solidification. This action negates the original strengthening mechanism provided by dispersion within the matrix. As a result, the weld metal will be significantly weaker than the base metal. Joining processes that do not involve melting of the base metal will retain the high strength of ODS alloys.

Nickel-base alloys are commonly welded by GTAW, SMAW, and GMAW. Frequently, a root pass is made by GTAW and the subsequent passes by GMAW. Submerged arc welding can be used on certain alloys, but the welding flux must be carefully selected to obtain adequate protection and provide correct elemental additions to the weld pool. The welding conditions chosen must avoid excessive heat input. When welding metal more than 75 mm (3 in.) thick, shrinkage stresses decrease ductility slightly, and a postweld stress-relieving treatment may be necessary. The manufacturer of the alloy should be consulted for specific details.

Precipitation-strengthened alloys are susceptible to cracking in the weld metal or in the HAZ, unless they are properly heat treated before and after welding. These alloys are normally welded in the solution annealed condition. If they are welded in the precipitation-strengthened condition, a solution anneal is required before high-temperature service. When solution annealing, a rapid heating rate should be used to avoid parent metal strain-age cracking. This usually can be accomplished by charging into a hot furnace. The alloys containing niobium and/or tantalum, such as Inconel 718, have a relatively slow hardening response and can be welded without undergoing spontaneous hardening during heating and cooling.

The nickel-chromium and nickel-molybdenum solid-solution alloys are readily welded in the annealed condition. No heat treatment is needed after welding to improve corrosion resistance, and generally alloys do not become embrittled after long exposure at temperatures up to about 815 °C (1500 °F). Table 13.1 lists general conditions for GTAW of superalloys.

Joint Design

The same joint designs are used for both GTAW and SMAW of nickel superalloys. Joint design for GMAW requires special consideration. Unless it has been proven satisfactory by experience, a joint design that has been developed for another metal should not be used for nickel-base superalloys.

Figure 13.2 shows joint designs used in welding nickel-base superalloys and gives the sizes of grooves or welds in different metal thicknesses and

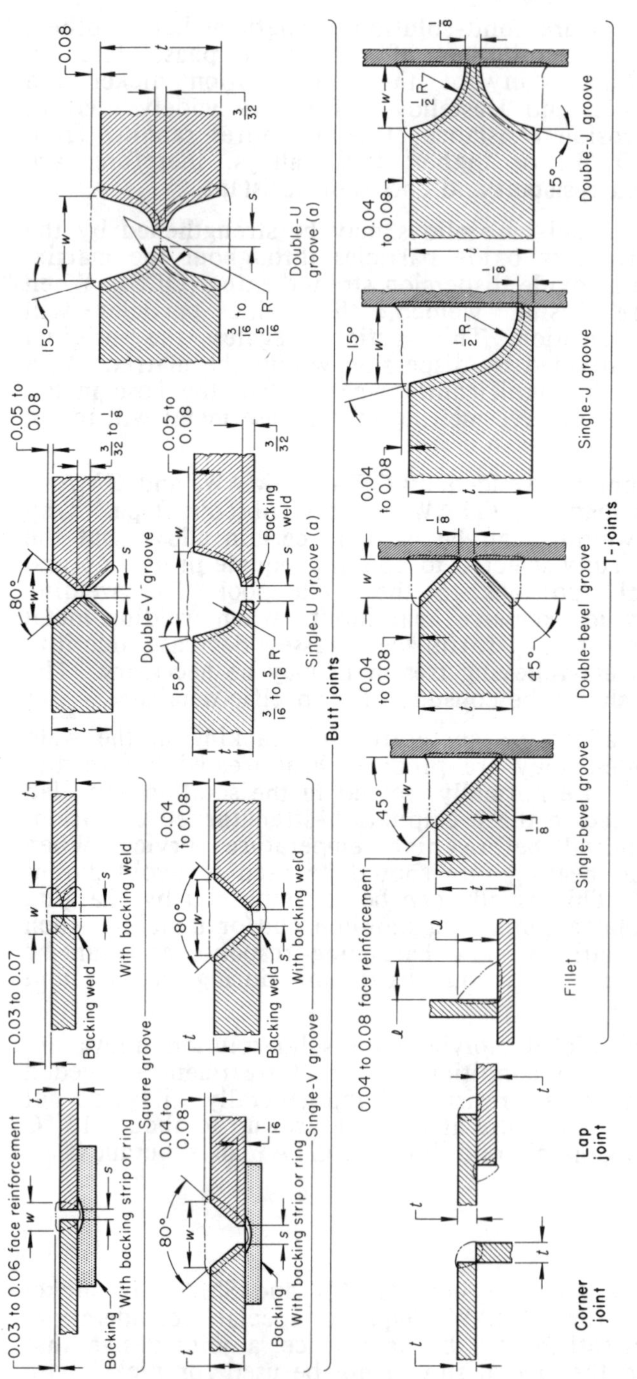

Fig. 13.2 Joint designs and dimensions for GTAW, GMAW, and SMAW

Base-metal thickness (t), in.	Width of groove or bead (w), in.	Maximum root opening (s), in.	Approximate amount of metal deposited lb/ft	Approximate weight of electrode, lb/ft(a)
Square-groove butt joint with backing strip or ring				
0.037	1/8	0	0.02	0.025
0.050	5/32	0	0.04	0.05
0.062	3/16	0	0.04	0.06
0.093	3/16-1/4	1/32	0.06	0.08
0.125	1/4	1/16	0.07	0.09
Square-groove butt joint with backing weld				
1/8	1/4	1/32	0.11	0.15
3/16	3/8	1/16	0.24	0.32
1/4	7/80	3/32	0.31	0.42
Single-V-groove butt joint with backing strip or ring				
3/16	0.35	1/8	0.227	0.31
1/4	0.51	3/16	0.443	0.61
5/16	0.61	3/16	0.582	0.80
3/8	0.71	3/16	0.745	1.02
1/2	0.91	3/16	1.16	1.59
5/8	1.16	3/16	1.61	2.21
Single-V-groove butt joint with backing weld				
1/4	0.41	3/32	0.42	0.58
5/16	0.51	3/32	0.54	0.74
3/8	0.65	1/8	0.73	1.00
1/2	0.85	1/8	1.21	1.67
5/8	1.06	1/8	1.46	2.00
Double-V-groove butt joint				
1/2	0.40	1/8	0.89	1.16
5/8	0.49	1/8	1.08	1.48
3/4	0.62	1/8	1.46	2.00
1	0.81	1/8	2.42	3.34
1 1/4	1.03	1/8	2.92	4.00
Single-U-groove butt joint(b)				
1/2	0.679	1/8	1.03	1.41
5/8	0.745	1/8	1.38	1.90
3/4	0.813	1/8	1.68	2.30
1	0.957	1/8	2.63	3.60
1 1/4	1.073	1/8	3.62	4.96
1 1/2	1.215	1/8	4.79	6.55
1 3/4	1.349	1/8	5.98	8.19
2	1.485	1/8	7.40	10.12
Double-U-groove butt joint(b)				
1	0.679	1/8	2.06	2.82
1 1/4	0.745	1/8	2.76	3.80
1 1/2	0.813	1/8	3.36	4.60
2	0.957	1/8	5.26	7.20
2 1/2	1.073	1/8	7.24	9.92
Corner and lap joint				
1/16	...	...	0.02	0.04
1/8	...	...	0.05	0.07
3/16	...	...	0.10	0.14
1/4	...	...	0.19	0.26
3/8	...	...	0.42	0.57
1/2	...	...	0.74	1.02
T-joint with fillet				
...	...	1/8	0.03	0.04
...	...	3/16	0.07	0.10
...	...	1/4	0.12	0.16
...	...	5/16	0.19	0.26
...	...	3/8	0.27	0.37
...	...	1/2	0.47	0.64
...	...	5/8	0.74	1.01
...	...	3/4	1.07	1.46
...	...	1	1.90	2.60

(continued)

Base-metal thickness (*t*), in.	Width of groove or bead (*w*), in.	Maximum root opening (*s*), in.	Approximate amount of metal deposited lb/ft	Approximate weight of electrode, lb/ft(a)
Single-bevel-groove T-joint				
1/4	0.125	...	0.07	0.09
5/16	0.188	...	0.13	0.17
3/8	0.250	...	0.19	0.26
1/2	0.375	...	0.38	0.52
5/8	0.500	...	0.63	0.86
3/4	0.625	...	0.93	1.28
1	0.875	...	1.77	2.42
Double-bevel-groove T-joint				
1/2	0.188	...	0.25	0.34
5/8	0.250	...	0.39	0.54
3/4	0.313	...	0.56	0.77
1	0.438	...	0.99	1.36
1 1/4	0.563	...	1.54	2.15
1 1/2	0.688	...	2.21	3.03
1 3/4	0.813	...	3.00	4.09
2	0.938	...	3.90	5.35
Single-J-groove T-joint				
1	0.625	...	1.78	2.4
1 1/4	0.719	...	2.50	3.4
1 1/2	0.781	...	3.23	4.4
1 3/4	0.875	...	4.09	5.6
2	0.969	...	4.93	6.8
2 1/4	1.031	...	5.80	8.0
2 1/2	1.094	...	6.94	9.5
Double-J-groove T-joint				
1	0.500	...	1.48	2.0
1 1/4	0.563	...	1.90	2.6
1 1/2	0.594	...	2.56	3.5
1 3/4	0.625	...	3.11	4.3
2	0.656	...	3.81	5.2
2 1/4	0.688	...	4.51	6.2
2 1/2	0.750	...	5.27	7.2

(a) To obtain linear feet of weld per pound of consumable electrode, take the reciprocal of pounds per linear foot. If the underside of the first bead is chipped out and welded, add 0.21 lb/ft of metal deposited (equivalent to 0.29 lb/ft of consumable electrode). (b) For GMAW (except with the short circuiting arc), root radius should be one half the value shown and bevel angle should be twice as great.

the approximate amount of metal deposited. Nickel-base alloy weld metal does not flow as readily or penetrate as deeply as steel weld metal does. Therefore, the joints must be more open to allow placement of weld metal, and lands should be thinner to accommodate lower penetration. Excessive puddling and heat input have a detrimental effect, because loss of residual deoxidizers may result. Use of the single- and double-bevel T-joint may not be suitable in some cases because of lack of joint accessibility.

Preweld and Postweld Heat and Mechanical Treatments

The solid-solution (non-age-strengthenable) alloys are welded in both the annealed and moderately cold worked conditions. The precipitation-strengthening alloys usually are welded in the solution treated condition. If a high degree of deformation should occur during preweld forming, or if the alloy has a high work-hardening rate, process annealing of the formed workpieces, before welding, may be required. Usually, preheating of nickel-base superalloys is neither needed nor recommended. A postweld thermal or mechanical treatment (e.g., shot peening) is sometimes needed, especially for the precipitation-strengthening alloys, to redistribute and relieve residual stresses resulting from weld-shrinkage strains. Often, both treatments are used on the same weldment.

Weldments made of solid-solution alloys can be used as-welded or after stress relieving, depending on the alloy and application. Stress relieving at temperatures ranging from 425 to 870 °C (800 to 1600 °F), depending on the alloy and its condition, can be used to reduce or remove stresses in work-hardened solid-solution alloys without producing a recrystallized grain structure. A low-temperature stress-equalizing heat treatment of 315 to 425 °C (600 to 800 °F) can be used to redistribute stresses without appreciably decreasing the mechanical strength produced by the previous cold working.

Precipitation-strengthening alloys are given a solution treatment after welding to relieve residual stresses, and then they are strengthened by an aging heat treatment. If the normal aging time or temperature is exceeded, overaging occurs; loss of strength and an increase in ductility can result.

For more information on joint design; preweld, postweld heat, and mechanical treatments; gas tungsten arc welding; gas metal arc welding; and shielded metal arc welding, see "Arc Welding of Heat-Resistant Alloys" in Volume 4 of the 9th Edition of *Metals Handbook.*

Causes and Prevention of Weld Defects

Weld defects such as cracks, porosity, inclusions, and incomplete fusion usually are unacceptable in weldments made of superalloys. Nondestructive inspection is used on almost all completed weldments; destructive inspection, of course, is limited to test samples or a sampling procedure on actual parts. Various types of leak tests are used on weldments that

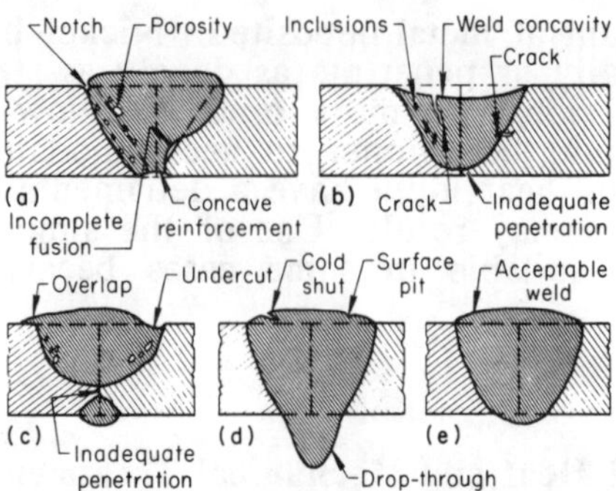

Superalloys (a-d). A weld with no defects and good reinforcement is shown in (e).

Fig. 13.3 Typical arc welding defects

are to be subjected to pressure in service. Typical types of defects that occur in arc welds are shown in Fig. 13.3.

Porosity and Inclusions

Porosity is the term used to describe gas pockets or voids in the weld metal. Typical causes include improper shielding, moisture, incorrect amperage, and excessive arc length. Dry electrodes are essential. When high amperage or a long arc length is used, deoxidizers that help prevent porosity can be totally consumed when transferring across the arc. Inclusions are usually slag, oxides, or other nonmetallic solids entrapped in the weld metal between adjacent beads or between the weld and parent metal. Excessive weld pool agitation, downhill welding, and undercutting can lead to slag entrapment. These conditions can usually be prevented by good weld practice and proper weld design.

Cracks and Fissures

The general types of cracks are: (*a*) transverse cracks in the base metal perpendicular to the weld; (*b*) longitudinal cracks in the base metal parallel to the weld; (*c*) microcracks and macrocracks in the weld metal; (*d*) centerline longitudinal weld-metal cracks; (*e*) crater cracks; and (*f*) start cracks or bridging cracks. Transverse cracks usually are the result of external contamination or a base metal with poor weldability. Base-metal cracking parallel to the weld typically is caused by the combination of a strong weld metal and weak, low-ductility base metal.

Weld-metal microfissuring can be caused by contamination or impurities in the metal that lower weldability. Centerline longitudinal cracking is caused by concave beads or a very deep, narrow weld bead. Crater cracking occurs when the arc is extinguished over a relatively large weld pool. The resulting concave crater is prone to shrinkage cracking.

Bridging cracks occur in highly stressed joints where good penetration is not achieved at the arc initiation point. Cracks of any type and size cannot usually be tolerated. If a material proves to be crack sensitive, base-metal cracking can be minimized by reducing heat input and depositing small beads, which results in lowered residual stresses.

Strain-Age Cracking

Strain-age cracking can occur in precipitation-strengthening nickel-base superalloys during the initial postweld heat treatment, if the base metal is in the aged condition and at least one area, such as the as-deposited weld metal, is not in the precipitation-treated condition. Cracks that form under these conditions are relatively large, and most of them are in the base metal. The metal is more likely to crack when an aged part is being repair welded. To minimize strain-age cracking:

* Weld in the solution treated (annealed) condition.
* Weld with minimum restraint.
* Do not preheat for welding.
* Use as low a heat input as possible.
* Accomplish postweld solution heat treatment by heating as rapidly as possible through the aging temperature range.

IRON-NICKEL-CHROMIUM AND IRON-CHROMIUM-NICKEL SUPERALLOYS

Iron-base superalloys include strain-hardenable, solid-solution-strengthened, and precipitation-strengthened types. All contain appreciable amounts of nickel and chromium, with either one or the other of these elements constituting the principal alloying addition. Other alloying elements generally are added to increase high-temperature strength (molybdenum, tungsten, and cobalt), to act as stabilizers (niobium and tantalum), or to promote strengthening (aluminum, titanium, and boron).

The usual range of service temperature for these alloys, 650 to 760 °C (1200 to 1400 °F), limits the selection of filler metals and preheat and postheat treatments for welding. Nominal compositions of the common iron-base alloys are given in Table 2.1, Chapter 2.

The solid-solution-strengthened alloys, such as N-155 and 16-25-6, are easily joined by SMAW, GMAW, and GTAW; however, the heat input should be kept low, and welds should be cooled rapidly to maintain ductility. Weld deposits can be made with an austenitic stainless steel filler metal, with a nickel-base alloy filler metal, or with a filler metal of the same composition as the base metal. Generally, preheating and postheating are used.

Some precipitation-strengthened alloys, such as A-286, are considerably more difficult to weld. These alloys are extremely sensitive to hot cracking in the weld metal and in the HAZ. Cracking is most likely to occur when aged metal or highly restrained parts are joined. Cracks in root passes or crater cracks can be minimized by using suitable welding procedures and techniques to control heat input during welding. Microcracking (microfissuring) can occur in the weld metal and in the HAZ. It must be controlled by proper preweld and postweld heat treatment and by selection of the most suitable filler metal.

Nevertheless, high joint efficiency can be obtained in arc welding A-286 sheet up to 2.4 mm (0.094 in.) thick. Joints in thicker sections are more difficult to weld and require special techniques such as automatic welding, a single pass, or the use of weld back cooling.

In all the precipitation-strengthened alloys, aluminum, niobium, or titanium may combine with either iron or the major solid-solution elements to form low-melting eutectic phases in the grain boundaries. Melting of these grain-boundary phases, often called incipient melting, occurs near the fusion line during welding, and thermal stresses incidental to weld cooling may cause the grains to separate. Grain separation produces gross subsurface cracks or microfissures that are difficult to detect by available nondestructive inspection methods.

Joint Design

Joint design for iron-base superalloys depends on the welding process, the application, and the filler metal used. For some applications, joint designs similar to those used for stainless steels are appropriate. For other applications, joint designs similar to those shown for nickel-base alloys are used (Fig. 13.2), except that the included angle of V-grooves is usually 60°. For SMAW, the included angle should be from 75 to 90°. Because nickel-base alloys have low fluidity, wider joint bevels and openings are sometimes required when welding with nickel-base filler metal.

Preweld and Postweld Heat Treatments

Strain-strengthened alloys usually are welded in the combined hot-cold worked condition. The metal in the HAZ is essentially solution treated by the welding heat, resulting in a decrease in hardness and strength. Although strain-hardenable alloys frequently are preheated for welding, the preweld heat treatments must be done in a limited temperature range to a avoid annealing the metal. This same restriction on heat treating temperature applies to postweld treatments of these alloys.

Solid-solution-strengthened alloys generally are welded in the solution treated condition and are used without postweld heat treatment. These alloys have a small zone of grain growth adjacent to the weld, but this does not appreciably reduce weld strength.

Precipitation-strengthened alloys are welded in the solution treated condition, because greater ductility of the base metal in this condition permits some relaxation of the shrinkage stresses associated with welding. Postweld heat treatment generally includes a re-solution treatment and an aging treatment.

For more information on gas tungsten arc welding, gas metal arc welding, shielded metal arc welding, and submerged arc welding, see "Arc Welding of Heat-Resistant Alloys" in Volume 4 of the 9th Edition of *Metals Handbook*.

COBALT SUPERALLOYS

Cobalt-base superalloys are available in both cast and wrought forms. Generally, the cast alloys are somewhat more difficult to weld than the wrought alloys, but they are still very weldable. Where the application requires very high reliability of welds, only GTAW and GMAW are recommended; otherwise, SMAW is used.

Joint Design

Joint design and weld grooves for cobalt-base alloys are essentially the same as for nickel-base alloys (Fig. 13.2). A square-groove butt joint is used for sheet metal up to about 2.8 to 3.2 mm (7/64 to 1/8 in.) thick, a V-groove for plate up to 9.5 mm (3/8 in.) thick, a double-V-groove or a double-U-groove for thicknesses of 9.5 to 15.9 mm (3/8 to 5/8 in.), and a double-U-groove for thicknesses over 15.9 mm (5/8 in.). Where T-joints are used, the same groove limitations apply as for butt joints. Corner joint welds should be backed by a fillet weld if possible. This type of joint should be avoided where high stresses are likely to occur. V-grooves should have a 60° groove angle for GTAW.

The weld grooves should be machined to ensure proper fit-up. The edges of a sheared plate should be ground or machined back 1.6 mm (1/16 in.) to remove stressed metal. Gas and arc cutting and beveling are not recommended. All joints should be designed to ensure full penetration.

Cleaning

The weld joint and adjacent area must be thoroughly cleaned before welding. All foreign matter should be removed by grinding, machining, or scrubbing with a suitable solvent. Shot or sand blasting should not be used, because iron and sand particles embedded in the work-metal surface can cause serious contamination. Wire brushing should be done with stainless steel. Copper brushes or carbon steel brushes can contaminate the base metal. If the alloys are heat treated in an oil-fired furnace, fuels of low sulfur content must be used to avoid sulfur contamination and subsequent detrimental effects on mechanical properties and corrosion resistance.

For more information on gas tungsten arc welding, gas metal arc welding, and shielded metal arc welding, see "Arc Welding of Heat-Resistant Alloys" in Volume 4 of the 9th Edition of *Metals Handbook.*

GAS METAL ARC WELDING OF SUPERALLOYS

Gas metal arc welding (GMAW), which often is called MIG (metal inert gas) welding, is an arc welding process in which the heat for welding is generated by an arc between a consumable electrode and the work metal. The electrode, a bare solid wire that is continuously fed to the weld area, becomes the filler metal as it is consumed. The electrode, weld pool, arc, and adjacent areas of the base metal are protected from atmospheric contamination by a gaseous shield provided by a stream of gas, or mixture of gases, fed through the welding gun. The gas shield must provide full protection, because even a small amount of air can contaminate the weld deposit.

Gas metal arc welding overcomes the restriction of using an electrode of limited length, as in shielded metal arc welding (SMAW), and overcomes the inability to weld in various positions, which is a limitation of submerged arc welding (SAW).

Gas metal arc welding is widely used in semiautomatic, machine, and automatic modes. In semiautomatic welding, the most popular method of applying this process, the welder guides the gun along the joint and adjusts the welding conditions. The wire feeder continuously feeds the filler-wire electrode, and the arc length is maintained by the power source. In automatic GMAW, the machinery controls the welding parameters, arc length, joint guidance, and wire feed, observed by the operator. Machine GMAW has only limited popularity, and it is characterized by machine control of the arc length, wire feed, and joint guidance. The operator adjusts the welding parameters.

Metals Welded

The nature of the gas metal arc process permits its use for welding most metals and alloys. However, because this book pertains only to superalloys, discussion of this welding process will concentrate on these alloys.

Solid-solution-strengthened nickel-base superalloys and, with suitable welding procedures, many precipitation-strengthened superalloys can be joined by GMAW. Gas metal arc welding is best suited to the joining of thick sections of more than about 6.4 mm ($^1/_4$ in.) thick, where high filler-metal deposition rates are desirable.

Spray, pulsed arc, globular, and short circuiting metal transfer can be used. Optimum metal transfer is obtained when operating slightly above the transition from globular to spray transfer. All of these methods use electrode wire of comparatively small diameter. Incomplete fusion and oxide inclusions can occur when the short circuiting arc is used. Multiple-pass welds should be made by highly skilled welders only. Direct current electrode positive (DCEP) should be used because the greater heating effect of reverse polarity assists in obtaining the required high melting rate.

Shielding Gas

Shielding gas for nickel-base superalloys is argon or an argon-helium mixture. Gas flow rates range from 10 to 30 L/min (25 to 60 ft^3/h), depending on joint design, type of metal transfer, and welding position. Generally, flow rate is about 25 L/min (50 ft^3/h) for spray transfer. As the percentage of helium is increased, gas flow rate must be increased to provide adequate protection. Pure argon is normally used for spray transfer. Other types of metal transfer commonly use argon with 25 to 30% helium added.

GAS TUNGSTEN ARC WELDING OF SUPERALLOYS

Gas tungsten arc welding (GTAW), often called TIG (tungsten inert gas) welding, is an arc welding process in which the heat is produced between a non-consumable electrode and the work metal. The electrode, weld pool, arc, and adjacent heated areas of the workpiece are protected from atmospheric contamination by a gaseous shield. This shield is provided by a stream of gas (usually an inert gas), or a mixture of gases. The gas shield must provide full protection; even a small amount of air can contaminate the weld.

The arc and weld pool are visible to the welder in GTAW. Slag that may be entrapped in the weld is not produced, and filler wire is not transferred across the arc, thus eliminating weld spatter. Because the electrode is nonconsumable, a weld can be made by fusion of the base metal without the addition of filler metal. A filler metal may be used, however, depending on the requirements that have been established for the particular joint.

Gas tungsten arc welding is an all-position welding process and is especially well adapted to the welding of thin metal - often as thin as 0.15 mm (0.005 in.). The process can be applied by the manual, semiautomatic, machine, or automatic methods. Manual welding is used for most GTAW applications.

Nickel-base superalloys are readily weldable by GTAW. This process is widely used for welding thin sections and for applications where a flux residue would be undesirable. Thin sections of aluminum-containing, precipitation-strengthened alloys are frequently joined without filler metal. The addition of filler metal is usually recommended for solid-solution alloys. Direct current electrode negative (DCEN) is recommended for both manual and automatic GTAW. Alternating current can be used for automatic current if the arc length can be closely controlled.

The welding arc is started by a high-frequency current. Extensions on the workpiece (start-up and run-off pads) that are machined off before the weldment is put into service are frequently used to ensure full-penetration welds and to minimize cracks in the weld metal caused by starts and stops. Heat input is kept as low as possible to minimize annealing and grain growth in the HAZ. General conditions for GTAW of nickel-base superalloys are shown in Table 13.1.

Shielding Gas

Argon, helium, or a mixture of argon and helium is used as shielding gas. The arc characteristics and heat pattern are affected by the choice of shielding gas. This choice should be based on welding trials for the particular production operation. Argon is normally used for manual welding; helium has shown some advantages over argon for machine welding thin sections without the addition of filler metal. Typical settings for welding 1.2-mm (0.045-in.) thick INCO 718 using argon and helium shielding are given in Table 13.2.

Welding-grade argon and helium should be used; oxygen, carbon dioxide, or nitrogen in the shielding gas are not used, because they reduce the service life of the tungsten electrode and can cause porosity in certain

Table 13.2 Typical settings for welding INCO 718

Operating condition	Shielding gas Argon	Helium
Current, A	80	40
Voltage, V	8-16	16-18
Welding speed, mm/s	3.4	2.5-3.4
(in./min)	(8)	(6-8)
Filler-wire diam, mm	0.75-0.89	0.75-0.89
(in.)	(0.030-0.035)	(0.030-0.035)
Wire-feed rate, mm/s	5-6.3	3.4-3.8
(in./min)	(12-15)	(8-9)
Torch gas flow, L/min	9.4-11.3	9.4
(ft^3/h)	(20-24)	(20)
Backing-gas flow rate, L/min	1.9	1.9
(ft^3/h)	(4)	(4)

alloys. An addition of about 5% hydrogen to argon acts as a reducing agent and is sometimes beneficial when the work metal has not been cleaned thoroughly. However, argon with 5% hydrogen should be used only for first-pass or single-pass welding, because porosity can result if this mixture is used for subsequent passes in multiple-pass welding.

Filler Metals

Filler metals used with nickel-base superalloys usually have the same general composition as the alloy being welded. Because of high arc currents and high welding temperatures, compositions of filler metals are often modified to resist porosity and hot cracking of the weld metal. Tack welding and root-pass welding without filler metal are permissible for some alloys. However, care must be taken to avoid centerline splitting and crater cracking when no filler metal is used. To minimize cracking, concave welds should be avoided. Table 13.3 gives the compositions of filler metals commonly used in GTAW; several of these filler metals are also used for welding metals other than nickel-base superalloys.

For welding the precipitation-strengthened nickel-base alloys, either a precipitation-strengthened or a solid-solution filler metal may be used, depending on service requirements. Maximum mechanical properties, particularly in thick metal, are obtained when precipitation-strengthened filler metals are used, because most of the weld deposit is composed of filler metal. The solid-solution filler metals produce welds with lower mechanical properties, but they can be used where maximum strength is not needed. For example, when welding Inconel 718 using either Rene′ 41, Inconel 718, GMR-235, Hastelloy S (AMS 5838), or Inconel 82 as the filler metal, weld specimens using the first three filler metals (precipitation strengthened) exhibit tensile properties similar to those of the base metal, but Hastelloy S and Inconel 82 filler metals (solid-solution strengthened) exhibit tensile properties about one third lower than those of the base metal.

Filler metal of the ERNiCr-3 classification (Table 13.3) is used for welding nickel-chromium-iron alloys to each other and to dissimilar metals, for high-temperature service, and for nuclear applications. Filler

Table 13.3 Compositions of filler metals and electrode wires for arc welding of superalloys

AWS classification or trade name	Composition, %												
	C	Mn	Fe	S	Si	Cu	Ni(a)	Co	Al	Ti	Cr	Nb + Ta	Mo	Other
Nickel-base bare electrodes for GTAW and GMAW														
ERNiCr-3	0.10	2.5-3.5	3.0	0.015	0.50	0.50	67 min	(b)	...	0.75	18.0-22.0	2.0-3.0(c)	...	0.50
ERNiCrFe-5	0.08	1.0	6.0-10.0	0.015	0.35	0.50	70 min	...	...	...	14.0-17.0	1.5-3.0	...	1.0
ERNiCrFe-6	0.08	2.0-2.7	10.0	0.015	0.35	0.50	67 min	...	...	2.5-3.5	14.0-17.0	...	...	0.50
ERNiCrFe-7	0.08	1.0	5.0-9.0	0.01	0.50	0.50	70 min	...	0.40-1.00	2.00-2.75	14.0-17.0	0.70-1.20	...	0.50
ERNiCrMo-3	0.10	0.5	5.0	0.015	0.5	...	rem	1.0	0.4	0.4	20.0-23.0	3.15-4.15	8.0-10.0	...
GMR-235	0.16	0.25	9.0-11.0	0.03	0.6	...	rem	2.5	1.75-2.25	2.25-2.75	14.0-17.0	...	4.5-6.5	0.009 B
ERNiCrMo-2	0.05-0.15	1.0	17.0-20.0	0.03	1.0	...	rem	0.5-2.5	...	...	20.5-23.0	...	8.0-10.0	0.2-1.0 W
Hastelloy S	0.01	0.2	1.0	0.005	0.20	...	67	...	0.2	...	15.5	...	15.5	0.009 B, 0.02 La
ERNiCrMo-7	0.007	0.50	1.5	0.005	0.04	...	65	1.0	...	...	16	...	15.5	...
Haynes 556	0.10	1.5	...	0.005	0.40	...	20	20	0.3	...	22	0.1	3	0.9 Ta, 0.2 N, 2.5 W
ERNiCrMo-4	0.01	0.5	5.5	0.005	0.04	...	62	1.2	...	...	16	...	16	3.5 W, 0.35 V
Inconel 601	0.05	0.5	14.1	0.007	0.25	0.25	60.5	...	1.35	...	23.0	...	...	...
Inconel 617	0.07	0.02	0.4	0.005	0.14	...	54	12.5	1.0	0.24	22	...	9	...
Inconel 718	0.08	0.35	rem	0.015	0.35	0.3	50-55	1.0	0.2-0.8	0.65-1.15	17.0-21.0	4.75-5.5	2.8-5.5	(d)
René 41 (AMS 5800)	0.12	0.1	5.0	0.015	0.5	...	rem	10.0-12.0	1.4-1.6	3.0-3.3	18.0-20.0	...	9.0-10.5	(e)
Waspaloy (AMS 5828C)	0.07	0.10	0.75	...	0.1	...	rem	13.5	1.4	3.0	19.75	...	4.45	(f)

(continued)

(a) Contains incidental cobalt. (b) Cobalt, 0.10% max, when specified. (c) Tantalum, 0.30% max, when specified. (d) Phosphorus, 0.015%; boron, 0.006%. (e) Boron, 0.01%; total of other elements, 0.003%. (f) Boron, 0.005%; zinc, 0.04%. (g) Cobalt, 0.12% max, when specified. (h) Tantalum, 0.30% max, when specified. (j) Vanadium, 0.60%; phosphorus, 0.04%; total of other elements, 0.50%. (k) Phosphorus, 0.04% max; tungsten, 1.25 to 1.75%. (m) Phosphorus, 0.040% max; tungsten, 2.00 to 3.00%. (n) Phosphorus, 0.02% max; boron, 0.0015 to 0.0022%

Table 13.3 (continued)

AWS classification or trade name	Composition, %													
	C	Mn	Fe	S	Si	Cu	Ni(a)	Co	Al	Ti	Cr	Nb + Ta	Mo	Other
Nickel-base covered electrodes for SMAW														
ENiCrFe-1	0.08	1.5	11.0	0.015	0.75	0.50	68 min(a)	...	...	...	13.0-17.0	1.5-4.0	...	0.50
ENiCrFe-2	0.10	1.0-3.5	6.0-12.0	0.020	0.75	0.50	rem	...	...	...	13.0-17.0	0.5-3.0	0.5-2.5	0.50
ENiCrFe-3	0.10	5.0-9.5	6.0-10.0	0.015	1.0	0.50	rem	(g)	...	1.0	13.0-17.0	1.0-2.5(h)	...	0.50
EniMo-1	0.12	1.0	4.0-7.0	0.030	1.0	...	rem	2.5	...	...	1.0	...	26.0-30.0	(j)
ENiMo-3	0.12	1.0	4.0-7.0	0.030	1.0	...	rem	2.5	...	...	2.5-5.5	...	23.0-27.0	(j)
ENiCrMo-3	0.10	0.5	5.0	0.015	0.50	...	rem	1.0(a)	0.40	0.40	20.0-23.0	3.15-4.15	8.0-10.0	...
ENiCrMo-2	0.10	0.5	18.5	0.005	0.5	...	47	1.5	...	...	22	...	9	0.005 B
ENiCrMo-7	0.007	0.5	1.5	0.005	0.10	...	65	1.0	...	...	16	...	15.5	...
ENiCrMo-4	0.01	0.5	5.5	0.005	0.04	...	62	1.2	...	...	16	...	16	3.5 W, 0.35 V
Inconel 117	0.01	0.6-1.4	1.7	0.008	0.50	0.20	52	12.0	0.2	...	23.5	0-0.5	9.0	...
Iron-nickel-chromium, iron-chromium-nickel, and cobalt-base heat-resistant alloy filler metals														
19-9 W (AMS 5782)	0.07-0.13	1.00-2.00	rem	0.030	1.00	0.50	8.00-9.50	...	...	0.10-0.30	19.0-22.0	1.00-1.30	0.35-0.65	(k)
Multimet (N-155) (AMS 5794)	0.1	1.00-2.00	rem	0.030	1.00	...	19.00-21.00	18.5-21.0	...	...	20.0-22.5	0.75-1.25	2.5-3.5	(m)
A-286 (AMS 5804)	0.04-0.05	1.25-1.35	rem	0.008	0.70	...	25	...	0.24-0.32	2.2	15	0.10-0.12	1.25	(n)
HS-25 or L-605 (AMS 5796)	0.10	1.5	3 max	...	10 max	...	10	rem	...	...	20	...	...	15 W
Haynes 188	0.10	0.6	1.5	0.005	0.35	...	22	39	...	...	22	...	...	14.5 W, 0.04 La

(a) Contains incidental cobalt. (b) Cobalt, 0.10% max, when specified. (c) Tantalum, 0.30% max, when specified. (d) Phosphorus, 0.015%; boron, 0.006%. (e) Boron, 0.01%; total of other elements, 0.003%. (f) Boron, 0.005%; zinc, 0.04%. (g) Cobalt, 0.12% max, when specified. (h) Tantalum, 0.30% max, when specified. (j) Vanadium, 0.60%; phosphorus, 0.04%; total of other elements, 0.50%. (k) Phosphorus, 0.04% max; tungsten, 1.25 to 1.75%. (m) Phosphorus, 0.040% max; tungsten, 2.00 to 3.00%. (n) Phosphorus, 0.02% max; boron, 0.0015 to 0.0022%

metal of the ERNiCrFe-5 classification is used to weld nickel-chromium-iron alloys and Inconel 600. The niobium-plus-tantalum content of these filler metals minimizes hot cracking in the weld when high stress is developed, as when welding thick metal.

Filler metal of the ERNiCrFe-6 classification is used for welding some combinations of dissimilar metals. The deposited weld metal responds to age-strengthening treatments. The age-strengthening response of this filler material is slight and does not exclude its use at 535 to 815 °C (1000 to 1500 °F).

Filler metal of the ERNiCrFe-7 classification contains aluminum, titanium, niobium, and tantalum and is used for welding the precipitation-strengthened alloys. The deposited filler metal responds to aging treatments. The weldment must be stress relieved prior to aging. Filler metals of the ERNiCrMo-3, ERNiCrMo-4, and ERNiCrMo-7 classifications are intended for welding the nickel-chromium-molybdenum alloys. Aerospace Material Specification (AMS) 5838 (Hastelloy S) is used for welding a variety of nickel-chromium, nickel-chromium-molybdenum, cobalt-base, and iron-base superalloys. It is well suited for dissimilar metal welding and exhibits excellent high-temperature stability.

The filler metals listed in Table 13.3 by trade name have no applicable American Welding Society (AWS) classifications, but most have AMS designations that are given in the table. These filler metals are primarily used for welding alloys of the same composition, although they are sometimes used for welding alloys of a different composition. For instance, Rene′ 41 and GMR-235 filler metals have been used to weld Inconel 718.

Welding Techniques

When filler metal is used, the hot end of the wire must be kept under the shielding gas, and wire diameter should be no larger than the work-metal thickness. Excessive turbulence in the molten weld pool must be avoided; otherwise, the deoxidizing elements will burn out.

To ensure a sound weld, the arc must be maintained at the shortest possible length. When no filler metal is added, arc length should not exceed 1.2 mm (0.05 in.) and preferably should be 0.5 to 0.75 mm (0.02 to 0.03 in.). When filler metal is added, the arc is longer, but it should be as short as possible, consistent with filler-metal diameter. Filler metals often contain elements specifically added to improve resistance to cracking and porosity. To obtain the full benefit of these elements, the finished weld should consist of about 50% filler metal.

A greater-than-normal electrode extension is needed for fillet welds and for the first few passes on heavy sections. Small-diameter filler-metal wires and more passes may be used on welds made in other than the flat position, for adequate control of weld metal. When the back sides of butt welds do not show adequate penetration, they should be ground back to sound metal and back beads should be deposited. When possible, backing gas should be provided when welding the first side.

SHIELDED METAL ARC WELDING OF SUPERALLOYS

Shielded metal arc welding (SMAW) is widely used for joining solid-solution nickel-base superalloys, but is rarely used for joining precipitation-strengthened superalloys. It is a process in which the heat for welding is generated by an arc established between a flux-covered consumable electrode and a workpiece. The electrode tip, molten weld pool, arc, and adjacent areas of the workpiece are protected from atmospheric contamination by a gaseous shield obtained from the combustion and decomposition of the electrode covering. Additional shielding is provided for the molten metal in the weld pool by a covering of molten flux or slag. Filler metal is supplied by the core of the consumable electrode and from metal powder mixed with the electrode covering of certain electrodes. Shielded metal arc welding is often referred to as arc welding with stick electrodes, manual metal arc welding, and stick welding. Direct current electrode positive is generally used to obtain optimum mechanical properties.

Weaving is sometimes desirable, but the amount should not exceed three times the electrode diameter. Overheating can cause hot short cracking in the weld metal or the base metal and excessive carbide precipitation at grain boundaries in the HAZ. This can be avoided by using the recommended amperage ranges, maintaining a short arc length, and not weaving the electrode excessively.

Electrodes

Electrodes used for shielded metal arc welding are listed in Table 13.3. Electrode composition should be similar to that of the base metal with which the electrode is to be used.

Welding Conditions

Figure 13.4 gives welding conditions for making butt, corner, and T-joints in solid-solution-strengthened nickel-base superalloys by SMAW. The weld metal of most nickel-base superalloys does not flow readily and must be manipulated or correctly positioned. This often requires a slight weave and a short pause at the sides to allow the undercut to be filled in. When the arc is broken, it should be shortened and the travel speed increased slightly. This reduces the weld pool size. When restarting, a reverse or T restrike should be used. The arc is struck at the leading edge of the weld crater and carried back to the rear of the crater. The travel direction is then reversed and normal weaving started.

All welding slag should be removed before placing a weld in service. Catastrophic high-temperature corrosion occurs if this is not done. Adhering slag can also enhance crevice corrosion at lower temperatures. Slag should also be removed between passes to ensure high-quality, metallurgically sound welds.

ELECTRON BEAM WELDING OF SUPERALLOYS

Electron beam welding (EBW) is a high-energy density fusion process that is accomplished by bombarding the joint to be welded with an intense (strongly focused) beam of electrons that have been accelerated up to velocities 0.3 to 0.7 times the speed of light at 25 to 200 kV, respectively. The instantaneous conversion of the kinetic energy of these electrons into

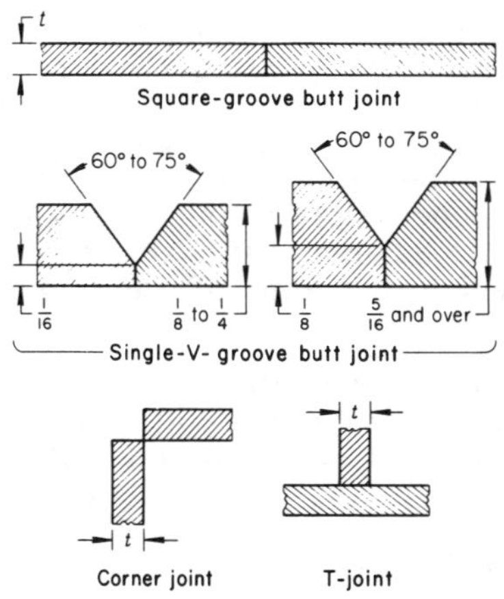

Metal thickness, in.	No. of passes	Current (DCEP), A(a)	Electrode diameter, in.(b)
Square-groove butt joints			
1/16	1	40-70	3/32
5/64	1	40-70	3/32
3/32	2	45-75	3/32
Single-V-groove butt joints			
1/8	2	40-70	3/32
5/32	2	40-70	3/32
3/16	2-3	40-70	3/32
1/4	3-4	40-130	3/32-5/32
3/8	5-6	40-130	3/32-5/16
1/2	8-10	40-130	3/32-5/16
Corner joints and T-joints(c)			
1/16	1	40-70	3/32
5/64	1	40-70	3/32
3/32	1	40-70	3/32
1/8	1	40-70	3/32
5/32	1	40-100	3/32-1/8
3/16	1	40-100	3/32-1/8
1/4	2	40-130	3/32-5/32
3/8	3	40-130	3/32-5/32
1/2	6	40-130	3/32-5/32

(a) Current should be within the range recommended by the electrode manufacturer. (b) Where a range is shown, the smaller diameters are used for the first pass in the bottom of the groove, and the larger diameters are used for the final passes. (c) Fillet welds

Fig. 13.4 Conditions for SMAW of solid-solution-strengthened nickel-base superalloys

thermal energy as they impact and penetrate into the workpiece on which they are impinging causes the weld-seam interface surfaces to melt and produces the weld-joint coalescence desired. Electron beam welding is used to weld any metal that can be arc welded; weld quality in most metals is equal to or superior to that produced by GTAW.

Because the total kinetic energy of the electrons can be concentrated onto a small area of the workpiece, power densities as high as 10^6 W/cm^2 can be achieved. That is higher than is possible with any other known continuous beam, including laser beams. The high-power density plus the extremely small intrinsic penetration of electrons in a solid workpiece result in almost instantaneous local melting and vaporization of the workpiece material. That characteristic distinguishes EBW from other welding methods in which the rate of melting is limited by thermal conduction.

Because of the marked differences in composition and weldability among nickel-, iron-, and cobalt-base superalloys, generalizations concerning electron beam welding of these alloys are not useful.

Solid-Solution Nickel Superalloys

Solid-solution nickel-base superalloys such as Hastelloy N, Hastelloy X, and Inconel 625 are readily electron beam welded. Hastelloy B and Inconel 600 can be welded to type 304 stainless steel and to themselves.

Precipitation-Strengthened Nickel Superalloys

Precipitation-strengthened nickel-base superalloys that are rated good in weldability by the electron beam process include Inconel 700, alloy 718, Inconel X-750, and Rene′ 41. Alloy 718 can be welded in either the annealed or the aged condition. Inconel X-750 should be welded in the annealed condition, and Rene′ 41 should be welded in the solution treated condition. Alloys of this group that have fair weldability include casting alloys INCO 713C and GMR-235 and wrought Udimet 700 and Waspaloy.

Iron-Nickel-Chromium Superalloys

Of the iron-nickel-chromium superalloys, alloy N-155 has good electron beam weldability, and alloys 16-25-6 and A-286 are rated fair. Alloy A-286 is usually welded in the solution treated condition; hot cracking may result if welded in the aged condition.

Cobalt Superalloys

Of the cobalt-base alloys, HS-21 has good weldability in unrestrained joints (and is generally poor in restrained joints). Cast alloy H-31 (X-40) has fair-to-good weldability, and alloy S-816 has fair weldability by the EBW process.

FRICTION WELDING OF SUPERALLOYS

Friction welding (FRW) is a process in which the heat for welding is produced by direct conversion of mechanical energy to thermal energy at the interface of the workpieces without the application of electrical energy, or heat from other sources, to the workpieces. Friction welds are made by holding a nonrotating workpiece in contact with a rotating workpiece under constant or gradually increasing pressure until the interface reaches welding temperature and then stopping rotation to complete the weld. The frictional heat developed at the interface rapidly raises the temperature of the workpieces, over a very short axial distance, to values approaching, but below, the melting range; welding occurs under the influence of a pressure that is applied while the heated zone is in the plastic temperature range.

Friction welding is classified as a solid-state welding process, in which joining occurs at a temperature below the melting point of the work metal. If incipient melting does occur, there is no evidence in the finished weld, because the metal is worked during the welding stage.

Most nickel-base and cobalt-base superalloys are easily friction welded to themselves and to alloy steels. The nickel-base superalloy GMR-235 can be welded to 1040 steel, Inconel 718 to Inconel 713C, and Inconel 713 to 8630 steel in producing jet engine parts that require high-strength bonds.

SOLID-STATE WELDING OF SUPERALLOYS

Solid-state welding (SSW) processes are those that produce coalescence at temperatures below the melting point of the base metal being joined. These processes involve either the use of deformation, or diffusion and limited deformation, to produce high-quality joints between both similar and dissimilar materials.

One form of SSW, called diffusion welding, is accomplished by bringing the surfaces to be welded together (faying surfaces) under moderate pressure and elevated temperature in a controlled atmosphere so that a coalescence of the interfaces or faying surfaces can occur. The other form, called deformation welding, is accomplished by subjecting the surfaces to be welded to extensive deformation. Melting or fusion is not associated with either process. Because diffusion welding and deformation welding both may be accomplished by the application of heat and pressure, some specific processes may share characteristics of both methods.

The driving force for the application of diffusion welding to nickel-base superalloys stems from their poor fusion weldability and needs of the aerospace industry to produce reliable welds for high-performance hardware. Materials such as Udimet 700, Inconel 718, and Inconel 600 have been successfully diffusion welded. However, all have low carbon contents coupled with the presence of carbide formers (chromium, titanium, and molybdenum). Because of this, organic surface contaminants can be reduced to stable carbides on the faying surfaces. Also, these materials have low interstitial solubility for oxygen coupled with the presence of stable oxide formers (chromium, aluminum, and titanium). This phenomenon renders the base metal highly susceptible to environmental contamination.

Because of the above reasons, great care must be exercised to ensure that weld surfaces are thoroughly cleaned before welding. It is also necessary to prevent recontamination during welding and to provide surface extension so that clean surfaces can come into intimate contact. Diffusion welding has not been widely applied to the superalloys. One limited application involves the manufacture of burner cans for the Pratt & Whitney TF30-P-100 military turbine engine. In this application, a hollow sandwich panel structure is made by diffusion welding two Hastelloy X face sheets to a corrugated center section. This structure is produced under the registered FINWALL trademark.

In general, diffusion welding of the iron-base and cobalt-base superalloys has not been pursued because of the ease with which they can be fusion welded.

Chapter 14

Brazing

INTRODUCTION

Brazing is a joining process that produces coalescence of materials by heating them to a suitable temperature and by using a filler metal that has a liquidus above 450 °C (840 °F) and below the solidus of the base metal. The filler metal is distributed between the closely fitted faying surfaces of the joint by capillary action. Joining with metals that melt at temperatures below 450 °C (840 °C) is classified as soldering. Soldering and low-temperature brazing, such as 650 °C (1200 °F) silver braze, however, are seldom used with superalloys, because components made of these materials usually operate at such elevated temperatures that these low-temperature joining operations are impractical. There are a number of different brazing processes used for superalloy parts. Almost any metal, as well as nonmetallics, can be brazed to superalloys, if it can withstand the heat of brazing.

BRAZING FILLER METALS

The American Welding Society (AWS) has classified several gold-, nickel-, and cobalt-base brazing filler metals that can be used for elevated temperature service (Table 14.1). In addition to these brazing filler metals, there are many that are not classified by AWS. The AWS classified brazing filler metals are suitable for high-temperature service; however, if the application is for temperatures above 980 °C (1800 °F) or in severe environments, the required brazing filler metal may not be listed in Table 14.1. It should be noted that for lower service temperatures, copper (BCu) and silver (BAg) brazing filler metals have been used for many successful applications.

Generally, superalloys are brazed with nickel- or cobalt-base alloys containing boron and/or silicon, which serve as melting-point depressants. In many commercial brazing filler metals, the levels are 2 to 3.5% B and 3 to 10% Si. Phosphorus is another effective melting-point depressant for nickel and is used in filler metals from 0.02 to 10%. It is also used where good flow is important in applications of low stress, where temperatures do not exceed 760 °C (1400 °F).

In addition to boron, silicon, and phosphorus, chromium is often present to provide oxidation and corrosion resistance. The amount may be as

Table 14.1 AWS brazing alloys for elevated temperature service

AWS classification	Composition, % Cr	B	Si	Fe	C	P	S	Al	Ti	Mn	Cu	Zr	Ni	Other elements total	Solidus, °F	Liquidus, °F	Brazing range, °F
Nickel-base alloy filler metals(a)																	
BNi-1	13.0-15.0	2.75-3.50	4.0-5.0	4.0-5.0	0.6-0.9	0.02	0.02	0.05	0.05	...	...	0.05	rem	0.50	1790	1900	1950-2200
BNi-1a	13.0-15.0	2.75-3.50	4.0-5.0	4.0-5.0	0.06	0.02	0.02	0.05	0.05	...	...	0.05	rem	0.50	1790	1970	1970-2200
BNi-2	6.0-8.0	2.75-3.50	4.0-5.0	2.5-3.5	0.06	0.02	0.02	0.05	0.05	...	...	0.05	rem	0.50	1780	1830	1850-2150
BNi-3	...	2.75-3.50	4.0-5.0	0.5	0.06	0.02	0.02	0.05	0.05	...	...	0.05	rem	0.50	1800	1900	1850-2150
BNi-4	...	1.5-2.2	3.0-4.0	1.5	0.06	0.02	0.02	0.05	0.05	...	...	0.05	rem	0.50	1800	1950	1850-2150
BNi-5	18.5-19.5	0.03	9.75-10.50	...	0.10	0.02	0.02	0.05	0.05	...	...	0.05	rem	0.50	1975	2075	2100-2200
BNi-6	...	...	...	...	0.10	10.0-12.0	0.02	0.05	0.05	...	...	0.05	rem	0.50	1610	1610	1700-2000
BNi-7	13.0-15.0	0.01	0.10	0.2	0.08	9.7-10.5	0.02	0.05	0.05	0.04	...	0.05	rem	0.50	1630	1630	1700-2000
BNi-8	...	...	6.0-8.0	...	0.10	0.02	0.02	0.05	0.05	21.5-24.5	4.0-5.0	0.05	rem	0.50	1800	1850	1850-2000

AWS classification	Composition, % Au	Cu	Pd	Ni	Other elements total	Solidus, °F	Liquidus, °F	Brazing range, °F
Precious metals								
BAu-1	37.0-38.0	rem	...	...	0.15	1815	1860	1860-2000
BAu-2	79.5-80.5	rem	...	...	0.15	1635	1635	1635-1850
BAu-3	34.5-35.5	rem	...	2.5-3.5	0.15	1785	1885	1885-1995
BAu-4	81.5-82.5	...	...	rem	0.15	1740	1740	1740-1840
BAu-5	29.5-30.5	...	33.5-34.5	35.5-36.5	0.15	2075	2130	2130-2250

AWS classification	Composition, % Cr	Ni	Si	W	Fe	B	C	P	S	Al	Ti	Zr	Co	Other elements total	Solidus, °F	Liquidus, °F	Brazing range, °F
Cobalt-base alloy filler metals																	
BCo-1	18.0-20.0	16.0-18.0	7.5-8.5	3.5-4.5	1.0	0.7-0.9	0.35-0.45	0.02	0.02	0.05	0.05	0.05	rem	0.50	2050	2100	2100-2250

(a) If determined, cobalt is 0.1% maximum unless otherwise specified.
Source: AWS 5.8-81

high as 20%, depending on the service conditions. Higher amounts, however, tend to lower joint strength.

Cobalt-base filler metals are used mainly for brazing cobalt-base superalloy components, such as first-stage turbine vanes for jet engines. Most cobalt-base filler metals are proprietary. In addition to containing boron and silicon, these alloys usually contain chromium, nickel, and tungsten to provide corrosion and oxidation resistance and to improve strength.

Product Forms

Available forms of AWS classified and proprietary brazing filler metals include wire, foil, tape, paste, and powder. The form used is dictated frequently by the application. If the filler metal required for a specific application is only available as a dry powder, then brazing aids such as cements and pastes are available to help position the brazing filler metal.

These products are offered as:

* Brazing filler-metal powders
* Brazing tapes and foils
* Brazing wires

SURFACE CLEANING

Cleaning of all surfaces that are involved in the formation of the desired brazed joint is necessary to achieve successful and repeatable brazed joints. All obstruction to wetting, flow, and diffusivity of the thermally induced molten brazing filler metals must be removed from both surfaces to be brazed prior to fit-up assembly. The presence of contaminants on one or both surfaces to be brazed may result in void formation, restricted or misdirected filler-metal flow, and contaminants included within the solidified brazed area, which reduces mechanical properties of the resulting brazed joint. Common contaminants are oils, greases, residual zyglo fluids, pigmented markings, residual casting or coring materials, and oxides formed either through previous thermal exposure or by exposure to contaminating environments.

The two basic types of cleaning are:

* Mechanical
* Chemical

PREPARATION

Certain superalloys, particularly nickel-base alloys containing high percentages of aluminum and titanium, may require a surface pretreatment to ensure maintenance of the cleaned surfaces. This surface pretreatment after cleaning is generally an electroplate of nickel, commonly referred to as nickel flashing. Thickness of the plate flashing is kept under 0.015 mm (0.0006 in.) for alloys with less than 4% Ti plus aluminum and 0.02 to 0.03 mm (0.0008 to 0.0012 in.) for alloys with greater than 4% Ti plus aluminum. This promotes wettability in the braze joint without seriously affecting the braze strength and other mechanical properties of the braze. The thickness of nickel plating may have to be increased as the brazing temperature is increased and as the time above 980 °C (1800 °F) is increased. Titanium and aluminum may diffuse to the surface of the nickel plating upon heating.

FIXTURING

One prerequisite to successful brazing that is often neglected is proper fixturing, when required for assembly and brazing of various components. One type of fixturing is classified as cold fixturing and is used primarily for assembly purposes. Another type of fixturing, hot fixtures, is used in the furnace for brazing.

Cold Fixtures

In most cases, cold fixtures are made of hot and cold rolled iron, stainless alloys, nonferrous alloys, and nonmetals, such as phenolics and micarta. These fixtures are used for assembly and tack welding details. They need not be massive or heavy, but should be sturdy enough to assemble components as required by design.

Hot Fixtures

Hot fixtures must have good stability at elevated temperatures and the ability to cool rapidly; metals are not generally stable enough to maintain tolerances during the brazing cycle. Therefore, ceramics, carbon, or graphite is used for hot fixturing.

CONTROLLED ATMOSPHERES

Controlled atmospheres (including vacuum) are used to prevent the formation of oxides during brazing and to reduce the oxides present so that the brazing filler metal can wet and flow on clean base metal. Controlled atmosphere brazing is used widely for the production of high-quality joints. Large tonnages of assemblies of a wide variety of base metals are mass produced by this process.

Controlled atmospheres are not intended to perform the primary cleaning operation for the removal of oxides, coatings, grease, oil, dirt, or other foreign materials from the parts to be brazed. All parts for brazing must be subjected to appropriate prebraze cleaning operations as dictated by the particular metals. Controlled atmospheres commonly are employed in furnace brazing; however, they may also be used with induction, resistance, infrared, laser, and electron beam brazing. In applications where a controlled atmosphere is used, postbraze cleaning is generally not necessary. In special cases, flux may be used with a controlled atmosphere (*a*) to prevent the formation of oxides of titanium and aluminum when brazing in a gaseous atmosphere, (*b*) to extend the useful life of the flux, and (*c*) to minimize postbraze cleaning. Fluxes should not be used in a vacuum environment.

The types of atmospheres currently used are:

* Pure dry hydrogen
* Inert gases
* Vacuum
* Metal-metal oxide equilibria in hydrogen and vacuum

BRAZING OF NICKEL SUPERALLOYS

In the selection of a brazing process for nickel superalloys, the characteristics of the alloy must be considered carefully. These superalloys include alloys that differ significantly in physical metallurgy, such as precipitation-strengthened versus solid-solution-strengthened alloys, and in process history, e.g., cast versus wrought. These characteristics can have a profound effect on brazeability.

Precipitation-strengthened alloys present several difficulties not normally encountered with solid-solution-strengthened alloys. The precipitation-strengthened alloys usually contain appreciable (greater than 1%) quantities of aluminum and titanium. The oxides of these elements are almost impossible to reduce in a controlled atmosphere (vacuum or hydrogen). Therefore, nickel plating or the use of a flux is necessary to obtain a surface that allows wetting by the filler metal.

Because these alloys are strengthened at temperatures of 535 to 815 °C (1000 to 1500 °F), brazing at or above these temperatures may alter alloy properties.

Liquid metal embrittlement is another difficulty encountered in brazing precipitation-strengthened alloys. Many nickel superalloys crack when subjected to tensile stresses in the presence of molten metals. This usually is confined to the silver-copper (BAg) filler metals. If precipitation-strengthened alloys are brazed in the solution treated and aged condition, residual stresses are often high enough to initiate cracking.

Cleanliness

Cleanliness, as in all metallurgical joining operations, is important when brazing nickel superalloys. Cleanliness of the base metal, filler metal, and flux (when used for induction brazing) and purity of the atmosphere should be as high as practical to achieve the required joint integrity. Elements that cause surface contamination or interfere with braze wetting or flow should be avoided in prebraze processing. All forms of surface contamination such as oils, chemical residues, scale, or other oxide products should be removed by using suitable cleaning procedures. The use of nickel-base filler metals offers some cost effectiveness in this regard, because nickel-base brazes are known to be self-fluxing and thus more forgiving to slight imperfections in cleanliness.

Attempting to braze over the refractory oxides of titanium and aluminum that may be present on precipitation-strengthened nickel superalloys must be avoided. Procedures to prevent or inhibit the formation of these oxides before and/or during brazing include special treatments of the surfaces to be joined or brazing in a highly controlled atmosphere.

Surface treatments include electrolytic nickel plating and reducing the oxides to metallic form.

Atmospheres

Dry, oxygen-free atmospheres that frequently are used include inert gases, reducing gases, and vacuum. The brazing atmosphere, whether gaseous or vacuum, should be free from harmful constituents such as sulfur, oxygen, and water vapor. When brazing in a gaseous atmosphere, monitoring of the water vapor content of the atmosphere as a function of dew point (the temperature at which a vapor starts to condense into a liquid) is common practice. A dew point of −51 °C (−60 °F) is average; however, a dew point of −62 °C (−80 °F) or below produces a better quality braze.

Stresses

During brazing, residual or applied tensile stress should be eliminated or minimized as much as possible. Also, inherent stresses present in precipitation-strengthened alloys may lead to stress-corrosion cracking. Stress relieving or annealing prior to brazing is recommended for all furnace, induction, or torch brazing. Brazing filler metals that melt below the annealing temperature are likely to cause stress-corrosion cracking of the base metal.

Thermal Cycles

Consideration must be given to the effect of the brazing thermal cycle on the base metal. Filler metals that are suitable for brazing nickel superalloys may require relatively high thermal cycles. This is particularly true for the filler-metal alloy systems most frequently used in brazing these alloys - the nickel-chromium-silicon or nickel-chromium-boron systems.

Solid-solution-strengthened nickel superalloys such as Inconel 600 may not be adversely affected by nickel braze filler-metal brazing temperatures of 1010 to 1230 °C (1850 to 2250 °F). Precipitation-strengthened alloys such as Inconel 718 may, however, display adverse property effects when exposed to brazing cycles higher than their normal solution heat treatment temperatures. Inconel 718, for example, is solution heat treated at 955 °C (1750 °F) for optimum stress-rupture life and ductility. Brazing temperatures of 1010 °C (1850 °F) or above result in grain growth and an attendant decrease in stress-rupture properties, which cannot be recovered by subsequent heat treatment. Some precipitation-strengthened nickel superalloys, however, may combine the solution heat treatment with a furnace brazing operation.

Consideration of base-metal property requirements for service enables selection of an appropriate braze alloy. Lower melting temperature - below 1035 °C (1900 °F) - braze filler metals are available within the nickel-base alloy family and within other braze filler-metal systems. (See Table 14.1.)

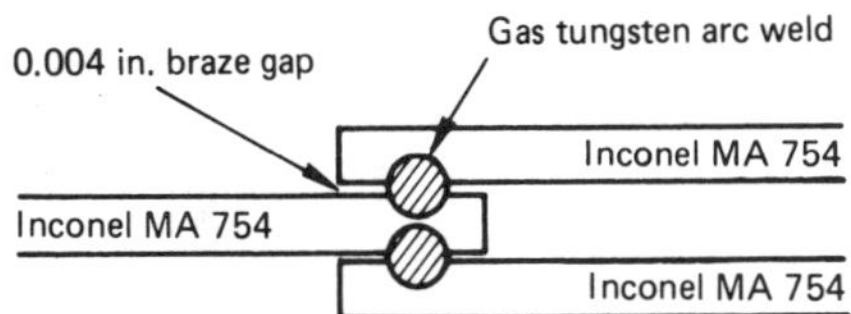

Fig. 14.1 Braze specimen assembly used for evaluating stress-rupture properties of brazed Inconel MA 754

Table 14.2 High-temperature brazement requirements of Inconel MA 754

Shear stress		Temperature		
MPa	ksi	°C	°F	Service life, h
26	3.8	980	1800	1000+
8.7	1.3	1095	2000	1000+

Inconel 718

Inconel 718 is often used in the fabrication of air diffusers for aerospace turbine engines. One manufacturer found that vacuum brazing of diffuser components at 10^{-4} torr in a cold-walled vacuum furnace provided the best results. Prior to brazing, all joint surfaces were nickel plated to 0.005- to 0.015-mm (0.0002- to 0.0006-in.) thicknesses. Plating was done in accordance with specification AMS 2424 or equivalent. Prior to assembly, application of BNi-2 braze filler-metal tape approximately 0.11 mm (0.0045 in.) thick was placed between all joint surfaces. After assembly, a braze slurry of BNi-2 filler metal was applied to all joints to ensure soundness.

BRAZING OF OXIDE-DISPERSION-STRENGTHENED ALLOYS

Oxide-dispersion-strengthened (ODS) alloys are powder metal alloys that contain stable oxides evenly distributed throughout the matrix. Oxide dispersion strengthening is used principally with nickel or nickel-chromium alloys, but iron-base alloys can also be dispersion strengthened. These oxides do not go into solution in the alloy at any time during the solid state. However, because the oxide is rejected from the matrix during melting and is not evenly redistributed on solidification, which occurs during fusion welding, these alloys are usually joined by brazing.

There are two types of ODS alloys: those that utilize only the oxide dispersoid for strengthening, and those that combine the dispersion strengthening with γ' precipitation strengthening. The mechanically alloyed (MA) alloys of INCO Ltd. are the predominant commercial ODS alloys.

Table 14.3 Typical compositions of several cobalt-base alloys

Alloy(a)	Nominal composition, % C	Mn	Si	Cr	Ni	Co	Mo	W	Nb	Ti	Al	B	Zr	Fe	Other	Characteristics and typical applications
AiResist 13(b)	0.45	0.5(c)	...	21	1.0(c)	rem	...	11	2.0	...	3.5	...	...	2.5(c)	0.1 Y	High-temperature parts
AiResist 213	0.18	...	...	19	...	rem	...	4.7	...	...	3.5	...	0.15	...	6.5 Ta, 0.1 Y	Sheets, tubing; resistant to hot corrosion
AiResist 215(b)	0.35	...	...	19	...	rem	...	4.5	...	...	4.3	...	0.13	...	7.5 Ta, 0.17 Y	Nozzle vanes; resistant to hot corrosion
Elgiloy	0.15	2.0	...	20.0	15.0	40.0	7.0	...	...	...	...	...	...	rem	0.04 Be	Springs; corrosion resistant, high strength
FSX-414(b)	0.25	1.0(c)	1.0(c)	29.5	10.5	rem	...	7.0	...	...	...	0.012	...	2.0(c)	...	Gas turbine vanes
FSX-418(b)	0.25	1.0(c)	1.0(c)	29.5	10.5	rem	...	7.0	...	...	...	0.012	...	2.0(c)	0.15 Y	Gas turbine vanes; improved oxidation resistance
FSX-430(b)	0.40	...	...	29.5	10.0	rem	...	7.5	...	...	...	0.027	0.9	...	0.5 Y	Gas turbine vanes; improved strength and ductility
X-40(b)	0.50	0.50	0.50	25	10	rem	...	7.5	...	...	...	...	...	1.5	...	Gas turbine parts, nozzle vanes
Haynes 150	0.08	0.65	0.35	28	3.0(b)	rem	1.5(b)	...	...	...	...	...	...	20.0	...	Resists thermal shock, high-temperature corrosion (air and air-SO_2)
Haynes 188 (sheet)	0.10	1.25(b)	0.3	22	22	rem	...	14	...	...	...	...	...	3.0(b)	0.04 La	Better oxidation resis-

MAR-M322(b) . . . 1.00	0.10	0.10	21.5	...	rem	...	9.0	...	0.75	...	...	2.25	...	4.5 Ta	Jet engine blades, vanes
MAR-M509(b) . . . 0.60	0.10(c)	0.10(c)	21.5	10	rem	...	7.0	...	0.2	...	0.010(c)	0.50	1.0	3.5 Ta	Jet engine blades, vanes
MAR-M918 . . . . 0.05	0.2(c)	0.2(c)	20	20	rem	...	...	...	...	...	...	0.16	0.5(c)	7.5 Ta	High-temperature sheets
MP35N	...	...	20.0	35.0	35.0	10.0	...	...	...	...	...	...	...	...	Stress corrosion resistant, high-strength fasteners
NASA Co-W-Re(b) . . . 0.40	...	...	3	...	rem	...	25	...	1.0	...	...	1.0	...	2.0 Re	High-temperature space applications
S-816 . . . . 0.38	1.20	0.40	20	20	rem	4.0	4.0	4.0	...	...	...	...	4	...	Gas turbine blades, bolts, springs
V-36 . . . . 0.27	1.00	0.40	25	20	rem	4.0	2.0	2.0	...	...	...	...	3	...	High-temperature sheets
WF-11, L605, Haynes 25 . . . 0.10	1.50	0.50	20	10	rem	...	15	...	...	...	...	...	3.0(b)	...	Jet engine parts, sheets
WF-31 . . . . 0.15	1.42	0.42	20	10	rem	2.6	10.7	...	1.0	...	...	...	...	...	High-temperature sheets
WI-52(b) 0.45	0.50(c)	0.50(c)	21	1.0(c)	rem	...	11	2.0	...	...	...	...	2.0	...	Gas turbine parts, nozzle vanes
X-45(b) 0.25	1.0(c)	...	25.5	10.5	rem	...	7.0	...	...	...	0.010	...	2.0(c)	...	Nozzle vanes

(a) Some superalloys are made by more than one manufacturer. The proprietary designation for such alloys has been used in this compilation. (b) Cast alloy. (c) Maximum composition

Nickel and Nickel-Chromium Alloys

Dispersion-strengthened nickel alloys and dispersion-strengthened nickel-chromium alloys, such as Inconel MA 754, are the easiest to braze of the ODS alloys. Vacuum, hydrogen, or inert atmospheres can be used for brazing. Prebraze cleaning consists of grinding or machining the faying surfaces and washing with a solvent that evaporates without leaving a residue. Generally, brazing temperatures should not exceed 1315 °C (2400 °F), unless demanded by a specific application that has been well examined and tested. The brazing filler metals for use with these ODS alloys usually are not classified by AWS. In most cases, the brazing filler metals used with these alloys have brazing temperatures in excess of 1230 °C (2250 °F). These include proprietary nickel-, cobalt-, gold-, or palladium-base alloys.

Brazements made of ODS alloys to be used at elevated temperature must be tested at elevated temperature to prove fitness-for-purpose. In the case of stress-rupture testing, AWS specification C3.2 may be used, because it represents the actual joint configuration. The configuration shown in Fig. 14.1 is preferred by some as a test model, although any test configuration without stress raisers is adequate. The elevated temperature brazement properties for Inconel alloy MA 754 should meet the requirements listed in Table 14.2.

Inconel MA 6000

Inconel MA 6000 is a nickel-base ODS alloy that is also γ' precipitation strengthened. The amount of alloying elements plus the γ' precipitation cause an interesting problem in the brazing of this alloy. Inconel MA 6000 has a solidus temperature of 1299 °C (2372 °F); therefore, the brazing temperature should be no higher than 1250 °C (2282 °F). Additionally, because 1230 °C (2250 °F) is the γ' solution treatment temperature, it becomes important to carefully select the brazing filler metal and to heat treat the assembly after brazing. The BNi, BCo, and specially formulated filler metals have been used for this alloy.

Inconel MA 6000 is used for its high-temperature strength and corrosion resistance; unfortunately, the passive oxide scale that provides good corrosion resistance also prevents wetting and flow of brazing filler metal. Therefore, correct cleaning procedures are very important. Surfaces to be brazed should be mechanically cleaned with a water-cooled, low-speed belt or wheel of approximately 320-grit and stored in a solvent, such as methanol, until immediately before the beginning of the brazing cycle.

BRAZING OF COBALT SUPERALLOYS

Brazing of cobalt superalloys is readily accomplished with the same techniques used for nickel-base alloys. Because most of the popular cobalt-base alloys do not contain appreciable amounts of aluminum or titanium, brazing atmosphere requirements are less stringent. Table 14.3 gives typical compositions of several cobalt-base alloys. These materials can be brazed in either a hydrogen atmosphere or a vacuum. Filler metals are usually nickel- or cobalt-base alloys or gold-palladium compositions. Silver or copper braze filler metals may not have sufficient

Table 14.4 Effect of brazing on mechanical properties of Haynes 25

Condition	Test temperature, °F	Ultimate strength, ksi	0.2% offset yield strength, ksi	Elongation, %
Tensile testing				
Mill anneal	Room	147.9	69.2	56
After braze cycle	Room	108.3	69.2	12
Mill anneal	1500	57.4	30.5	17
After braze cycle	1500	57.0	33.2	24
Mill anneal	1800	21.0	18.1	35
After braze cycle	1800	21.5	18.8	35

Condition	Test temperature, °F	Stress, ksi	Hours to rupture
Stress-rupture testing			
Mill anneal	1500	24.5	82
After braze cycle	1500	24.5	72.3
Mill anneal	1650	15.0	56
After braze cycle	1650	15.0	36
Mill anneal	1800	6.5	110
After braze cycle	1800	6.5	120

strength and oxidation resistance in many high-temperature applications. Although cobalt-base alloys do not contain appreciable amounts of aluminum or titanium, an electroplate or flash of nickel is often used to promote better wetting of the brazing filler metal.

Nickel-base brazing alloys such as AWS BNi-3 have been used successfully on Haynes 25 for honeycomb structures. It has been reported that after brazing a diffusion cycle is used to raise the braze joint remelt temperature to 1260 to 1315 °C (2300 to 2400 °F). Table 14.4 presents the effects of a high-temperature braze at 1225 °C (2240 °F) for 15 min on the mechanical properties of Haynes 25. One cobalt-base brazing filler metal (AWS BCo-1, Table 14.1) appears to offer a good combination of strength, oxidation resistance, and remelt temperature for use on Haynes 25 foil.

Cobalt alloys, much like nickel alloys, can be subject to liquid metal embrittlement or stress-corrosion cracking when brazed under residual or dynamic stresses. This frequently is observed when using silver or silver-copper (BAg) filler metals. Liquid metal embrittlement of cobalt-base alloys by copper (BCu) filler metals occurs with or without the application of stress; therefore, BCu filler metals should be avoided when brazing cobalt alloys.

ISOTHERMAL SOLIDIFICATION JOINING PROCESSES

New superalloy joining processes that are characterized by isothermal solidification have recently been developed. These processes are also known as transient liquid-phase techniques. Isothermal solidification joining is essentially a combination of diffusion welding and brazing. Diffusion welding is a nonfusion process that relies on careful surface preparation, the possible use of an interlayer material, and the application of pressure at the joint during the heating cycle to produce high-quality welds that match superalloy base-metal properties. The isothermal solidification joining process relies on the diffusion of constituents to facilitate joining by locally lowering the melting points of the interface metals. Continued diffusion during isothermal heating raises the melting point to produce joints that match base-metal properties.

In these isothermal solidification processes, boron often is used to depress the melting point of the joining composition. As the components are held at temperature and under pressure, diffusion of the depressant element occurs gradually, raising the melting point of the interface metal above the holding temperature. Thus, solidification occurs at constant temperature. A limitation of these processes is that boron can combine with elements such as chromium, thus forming borides rather than diffusing throughout the component in solution at very low concentrations. This results in a plane of weakness along the joint.

Chapter 15

Protective Coatings

INTRODUCTION

Protective coating development for superalloys and refractory metals has been spurred by advances in propulsion technology for aircraft and space vehicles since about 1960. These advances have placed increasing temperature and structural demands on materials for service at temperatures of 1010 °C (1850 °F) and above. This, coupled with weight-associated penalties for flight systems, has driven materials and design technology to maximize the hot strength of structural components.

To achieve the high-temperature mechanical properties demanded by modern gas turbine technology, superalloys with reduced chromium contents have had to sacrifice the oxidation and corrosion resistance that was inherent in previously used heat-resistant alloys with higher chromium contents. Consequently, coatings are relied on to protect superalloy components such as turbine blades and vanes from environmental attack. This advanced technology was initially required for advanced military aircraft gas turbines. The performance and fuel economy advantages associated with advanced turbine technology proved sufficiently attractive to cause advanced materials technology application, including the use of coatings, to be used in commercial aircraft, marine, and stationary turbines.

For superalloy metal substrates, the exclusive purpose of the coating is to prevent corrosive attack of the substrate for the maximum possible time with the maximum degree of reliability. Figure 15.1 illustrates the benefits derived from use of protective coatings at elevated temperatures. Surface corrosion resistance is increased significantly, as illustrated. Coatings that are used for protection against environmental attack are not in equilibrium with the substrate. At the high temperatures involved, interdiffusion between coating and substrate occurs, although at a relatively low rate in well-balanced coating/substrate systems.

Coatings fail eventually because of this interdiffusion, which causes the coating chemistry to change substantially so that it is no longer protective. Mechanically, physically, and chemically, coatings and substrates differ. In the design, development, and application of coated systems, the primary objective is not only to develop a coating material that provides corrosion protection, but to create a coated hardware system that provides optimum properties and surface stability for the intended service.

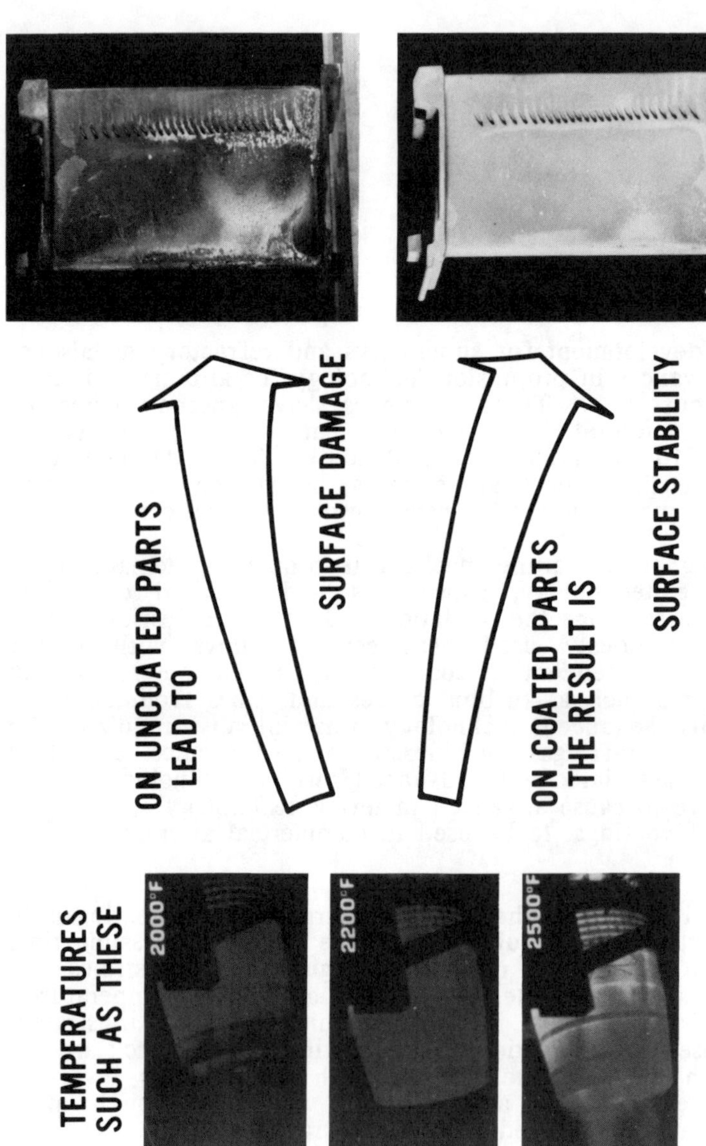

Fig. 15.1 Comparison of surface corrosion resistance of coated and uncoated superalloy parts

ENVIRONMENTAL EFFECTS

There are two general types of environmental effects that directly affect superalloys. These are oxidation and hot corrosion. Both processes involve rapid attack of the surface of superalloy parts and a resulting decrease in material stability. As such, coatings are designed to resist the specific effects of each process.

Oxidation

Superalloys generally react with oxygen, which is the primary environmental factor that affects service life. At moderate temperatures of about 870 °C (1600 °F) and below, general uniform oxidation is not a major problem. At higher temperatures, commercial nickel- and cobalt-base superalloys are attacked by oxygen. The level of oxidation resistance at temperatures below about 980 °C (1800 °F) is a function of chromium content - Cr_2O_3 forms as a protective oxide. At temperatures above about 980 °C (1800 °F), aluminum content becomes more important in oxidation resistance - Al_2O_3 forms as a protective oxide. Chromium and aluminum can contribute in an interactive fashion to oxidation protection. The higher the chromium level, the less aluminum may be required to form a highly protective Al_2O_3 layer. However, the aluminum contents of many superalloys are insufficient to provide long-term Al_2O_3 protection, and protective coatings consequently are used to provide satisfactory service life.

These coatings also prevent selective attack that occurs along grain boundaries and at surface carbides (Fig. 15.2). They also inhibit internal oxidation or subsurface interaction of O_2/N_2 with gamma prime (γ') envelopes, a process believed to occur in nickel-base superalloys.

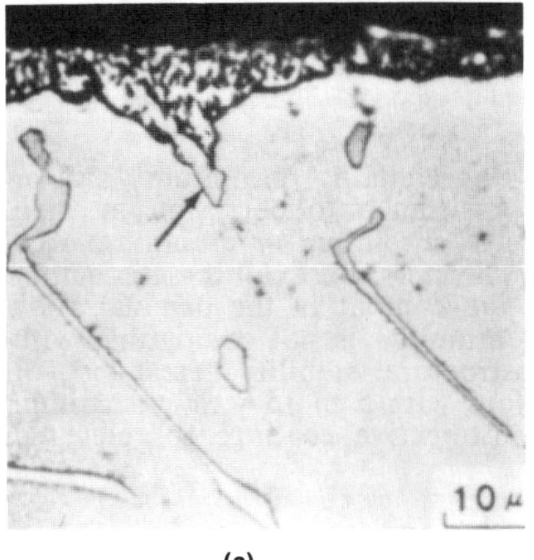

(a)

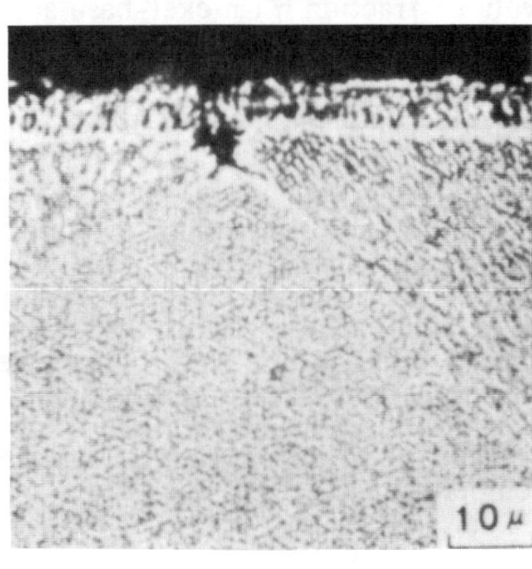

(b)

(a) Accelerated oxidation of MC carbide (arrow) at surface of MAR-M 200 at 925 °C (1700 °F). (b) Accelerated oxidation of grain boundary in Udimet 700 at 760 °C (1400 °F). Magnification, 1000×.

Fig. 15.2 Effects of oxidation on superalloys

Hot Corrosion

In lower temperature operating conditions, such as ≤870 °C (≤1600 °F), accelerated oxidation may occur in superalloys through the operation of selective fluxing agents. One of the most well-known accelerated oxidation processes is hot corrosion (sometimes known as sulfidation). The hot corrosion process is separated into two regimes: low temperature and high temperature. High-temperature attack occurs at 900 to 1050 °C (1650 to 1920 °F), and low-temperature attack occurs at 680 to 750 °C (1255 to 1380 °F). Hot corrosion is triggered by the presence of sulfur in fuel and impurities, particularly salt, in the environment.

The principal method for combating hot corrosion is the use of a high chromium content (≥20 wt%) in the base alloy. Although cobalt-base superalloys and many iron-nickel alloys have chromium levels in this range, many nickel-base alloys - especially those with high creep-rupture strengths at high temperatures - do not, because a high chromium content is not compatible with the high volume fraction (V_f) of γ' required. As chromium is increased in these alloys, the γ' solid solubility temperature is decreased, so that there is a decreased amount of γ' available for strengthening at high temperatures.

Higher titanium-aluminum ratios also seem to reduce attack on uncoated superalloys. Alloys with improved resistance to hot corrosion, based on slightly increased chromium contents and appropriate titanium-aluminum modifications, have been produced. For maximum uncoated hot corrosion resistance, however, chromium contents in excess of 20 wt% appear to be required. Such alloys are not capable of achieving the strengths of the high volume fraction γ' alloys such as MAR-M 200 and B-1900. Consequently, coatings that protect the base metal (overlay coatings seem to provide the best surface protection), or sometimes environmental inhibitors, are used to suppress hot corrosion attack in high-strength (high volume fraction γ') nickel-base superalloys.

COATINGS FOR SUPERALLOYS

Advanced, high-strength superalloys do contain chromium and/or aluminum additions that provide some resistance to corrosion at high temperatures by the formation of protective chromium or aluminum-rich oxides. For either chromia or alumina formers to exhibit satisfactory lives under cyclic conditions, the chromium content in the original alloy should be about 20 wt%. This level of chromium is not compatible with the high-temperature strength and microstructural stability demanded for the most rigorous service conditions. Alloys with 5 to 15 wt% chromium, typical of advanced superalloys, require protective coatings for elevated temperature use.

Coating Requirements

Gas turbine applications have proved to be the major impetus of coating technology for superalloys, because both oxidation and hot corrosion are involved. Hot corrosion and oxidation involve rapid attack and consumption of hardware, and coatings are designed to resist the specific attack anticipated for a particular turbine application. Marine propulsion turbines, for example, are susceptible to both forms of hot corrosion due to the high sulfur content of the fuel and the sea-salt service environment. Aircraft gas turbines, on the other hand, are more susceptible to oxidation, and sometimes high-temperature hot corrosion, if the aircraft operates consistently in a coastal or marine environment.

In addition to corrosion resistance, coatings for superalloys are required to resist both thermal cycling, often at rapid rates, and mechanical forces that act upon hardware, without cracking. The protective oxide film (alumina), on which coatings for superalloys rely for protection, does not readily span even the smallest of gaps. Cracking of protective coatings is soon followed by a breakdown in protection, and localized substrate attack follows shortly. Close correlation of thermal expansion properties between coating and substrate is important, and coatings with a small amount of ductility are desirable from a corrosion standpoint. Coatings also must be able to withstand combustion char and solid particulates ingested during turbine operations that result in erosive action.

Coatings must be reasonably compatible with the substrates to which they are applied. Components of the coating and the method of coating application should be selected to avoid undesirable reaction phases between coating and substrate and rapid penetration of the substrate by coating elements. The presence of such phases and/or interdiffusion leads to void formation, or cracking at the interface, and coating spall. These imperfections compromise the mechanical performance capability of the system and the protection it provides.

TYPES OF COATINGS

As a result of many years of development and use in government and industry, two basic types of coatings for superalloys have emerged. These are diffusion coatings and overlay coatings. Both types depend on the sacrificial oxidation of aluminum in the coating to provide a spall-resistant, protective alumina coating to prevent or strongly inhibit further attack.

Diffusion Coatings

In the diffusion coating application process, aluminum is made to react at the surface of the substrate, forming a layer of monoaluminide (known generally as MAl). For coatings applied over nickel-base superalloys, nickel aluminide (NiAl) is the resulting species, and over cobalt, it is cobalt aluminide (CoAl). In effect, diffusion coatings are surface conversion processes. This type of coating is modified to some extent by the elements contained in the substrate as the diffusion reaction proceeds and usually is further modified by other metallic elements that are intentionally added during the coating process to improve coating and/or system performance for specific service conditions.

Overlay Coatings

Overlay coatings do not rely on reaction with the substrate for their formation, although some moderate interdiffusion usually occurs during

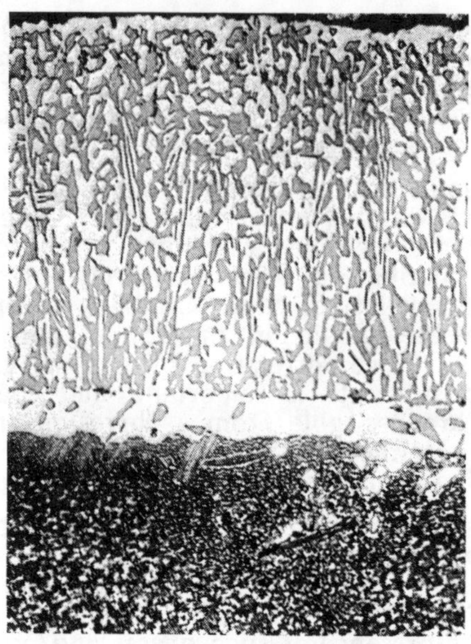

Source: "Protective Coatings for High Temperature Alloys: State of the Technology," written by G. William Goward, *Proceedings of the Symposium on Properties of High Temperature Alloys,* 1977, p 806. This figure was originally presented at the 1976 Fall Meeting of the Electrochemical Society, Inc. held in Las Vegas, Nevada.

Fig. 15.3 Microstructure of two-phase cobalt-chromium-aluminum-yttrium (CoCrAlY) overlay coating

service. Rather, the material applied over the substrate during the coating process and the specific method of processing determine the composition and microstructure of the coating. (See Fig. 15.3.) Coatings of this type in current use are generically called MCrAlY coatings and essentially comprise a monoaluminide (MAl) component contained in a more ductile matrix of solid solution (γ) in the case of cobalt-chromium-aluminum-yttrium (CoCrAlY) coatings, or a mixture of γ and γ' (Ni_3Al) phase in the case of nickel-chromium-aluminum-yttrium (NiCrAlY) coatings. Additionally nickel-cobalt-chromium-aluminum-yttrium (NiCoCrAlY) and iron-chromium-aluminum-yttrium (FeCrAlY) modifications are also available.

Service Characteristics

Coatings for superalloys, whether diffusion or overlay types, are applied with thicknesses of 50 to 150 μm (2 to 6 mils). Protective life increases in proportion to thickness. Thicker coatings in a given system are more prone to cracking in response to thermal cyclic-induced stresses. Thus, a hot military aircraft engine that is required to provide rapid acceleration and consequent rapid and extreme temperature changes in the turbine section may preclude the use of thicker coatings. Frequent inspection and short time between overhauls (TBO) of 100 to 500 h may be needed.

Engines for commercial aircraft, some military aircraft, and power generation service may be designed for more moderate temperature and loading conditions, and protective coatings can approach the higher end of the thickness range with less risk of damage due to cracking and spalling. Longer lifetimes are achieved, and the time between overhauls can run into thousands of hours. During overhaul, turbine blades and vanes that have not exceeded creep limits and are not otherwise severely eroded or damaged are refurbished for reuse. Coatings are stripped. The parts are reworked and cleaned as necessary, recoated, and returned to service. Some coatings need not be completely stripped before recoating, if weld repair is not required.

METHODS OF APPLYING DIFFUSION COATINGS

Many organizations have the capacity and facilities to apply diffusion coatings to superalloys. These include the manufacturers of gas turbine engines, suppliers of turbine hardware, specialty coating houses, and users such as military aircraft maintenance facilities and commercial airlines. Two basic methods for the application of diffusion coatings are used in these operations, but many variations of these methods exist.

The two basic methods used for aluminizing superalloys are slurry fusion and pack cementation. In the slurry process, the part is sprayed with a suspension of aluminum or aluminum alloy and is subjected to a high-temperature treatment to produce melting and interdiffusion between deposit and substrate. In the pack cementation process, the part is reacted with aluminum or aluminum alloy powder in the presence of an ammonium halide activator at an elevated temperature. The operation is carried out in a hermetically sealed container to maintain controlled activity of the aluminum in the vapor phase. Advanced coating treatments include additions of chromium, platinum, or rhodium. Improvements in hot corrosion resistance are achieved by proper control of the platinum distributions in the aluminide layer. The platinum-rich coatings are synthesized by electroplating a thin platinum layer before aluminizing.

Some aspects of most processes for coating superalloy components are proprietary to the coating organization. Specific processes, and sometimes operating personnel, must be properly certified for processing gas turbine components.

METHODS OF APPLYING OVERLAY COATINGS

Three methods have been developed for applying overlay coatings to superalloys. These are:

* Physical vapor deposition (PVD)
* Plasma spraying
* Sputtering

Of these methods, two are used routinely. Use of PVD by melting and evaporating a coating material source bar with a focused electron beam in an evacuated chamber is one commercial method. Plasma spraying of prealloyed powders is the other. Sputtered overlay coatings of good quality can be achieved, but this method of application is not yet used commercially.

All current methods of applying overlay coatings are limited to line-of-sight application. Only surfaces directly exposed to the flow of material emanating from the plasma torch or vapor flux can be coated with acceptable uniformity. Vapor phase diffusion coatings, by contrast, allow internal passages and recessed details to be coated effectively.

For more detailed information on diffusion and overlay coatings, see "Oxidation Protective Coatings for Superalloys and Refractory Metals" in Volume 5 of the 9th Edition of *Metals Handbook*.

Chapter 16

Identification, Recycling

INTRODUCTION

Quality assurance during the fabrication of hardware, subassemblies, and assemblies sometimes requires a reliable system of rapid identification of metals and alloys. Because various metals may become mixed during storage or use, and because lengths of strip, sheet, plate, billets, bar, wire, and fabricated products may have lost their identifying marks, some means of sorting mixed lots is necessary. In addition, it is equally important to have a means of quickly identifying superalloys, or other materials that are supposed to be superalloys, during scrap operations.

Because of the high cost and/or periodic scarcity of superalloys, recycling of scrap is used extensively for their recovery. A survey of the industry in 1976 showed that the superalloys produced in the United States were melted using an average of 42% "home" (in-house) scrap, 17% purchased scrap, and 41% primary metal.

Traditionally, the superalloy industry has required that the highest quality raw materials be used to begin the melt, because the furnaces used have had limited refining capacity, and the highly critical nature of many parts manufactured from superalloys has demanded high-purity starting materials. However, the innovation of the argon-oxygen decarburization (AOD) process has provided a means of removing many impurities, thus permitting use of a much lower grade of scrap. Other new systems, including the vacuum-oxygen decarburization (VOD) process and plasma-arc refining, offer additional refining capabilities for producing high-grade superalloys from lower grades of scrap. The criticality of some parts such as aircraft engine disks, however, still requires that the highest quality raw materials be used.

Progress also is being made in segregating and identifying superalloy turnings. "Pedigree" turnings are routinely degreased, fragmented, and compressed for remelting. Considerable attention currently is being directed toward the recovery of superalloys from grindings, although they are generally lower in purity than turnings and thus offer more problems in reclamation.

MEANS OF IDENTIFICATION

The best way of identifying metals is by quantitative chemical analysis. However, chemical or spectrographic analyses require extensive, time-consuming procedures and expensive equipment that may not be fully utilized. Also, a complete chemical analysis often may be unnecessary.

Common methods of rapid identification of metals include techniques involving:

* Color
* Weight
* Magnetic properties
* Spark testing
* Chemical spot testing

Each of these topics is discussed briefly below. For more detailed information, see Section 31, "Recycling of Metals and Alloys," and Section 33, "Rapid Identification of Metals and Alloys," in *Metals Handbook; Desk Edition.*

Color

Many metals and alloys have a characteristic color that can be utilized for initial separation. A very preliminary sorting based on color can be carried out according to Table 16.1.

Table 16.1 Preliminary identification of metals and alloys by color

Color	Metal or alloy
Red or reddish	Copper
Light brown or tan	90/10 cupronickel
Dark yellow	Bronzes, gold
Light yellow	Brasses
Bluish or dark gray	Lead, zinc, zinc alloys
White or light gray	Nearly all others

Weight

Various metals may vary greatly in specific gravity - for example, from 1.74 for magnesium to 21.4 for platinum. Thus, it may be possible to make an initial separation, classifying them into categories, as shown in Table 16.2.

Table 16.2 Preliminary identification of metals by weight

Weight	Metals	Specific gravity
Very heavy	Gold, platinum group, tungsten	19.3-21.4
Heavy	Lead, silver, molybdenum	10.2-11.3
Light	Magnesium, aluminum, titanium	1.7-4.5
Intermediate	Nearly all others	6-9

MAGNETIC TESTING

Strongly magnetic metals include gray, ductile, and malleable irons; carbon steels; alloy steels; tool steels; ferritic and martensitic stainless steels; and high-nickel alloys. Other strongly magnetic metals are the iron-silicon alloys containing 0.5 to 4.5% Si; iron-nickel alloys, particularly those containing more than 28% Ni; iron-cobalt alloys; and iron-molybdenum alloys. In differentiating among these strongly magnetic materials, it should be noted that the high-nickel alloys such as superalloys are less magnetic than the magnetic iron-base materials. Table 16.3 lists typical magnetic responses of various metals that can be used in the preliminary identification of scrap materials.

Table 16.3 Preliminary identification of metals and alloys by magnetic response

Response	Metal or alloy
Strongly magnetic	Cast irons, steels, 400 stainless steels, nickel, cobalt
Slightly magnetic	Monel (not K or S Monel), aluminum bronze, manganese bronze, silicon bronze
Nonmagnetic	Nearly all others

SPARK TESTING

Spark testing is used to classify ferrous alloys according to their chemical compositions, by visual examination of the spark pattern or stream that is thrown off when the alloys are held against a grinding wheel rotating at high speed. The test offers a fast and economical means of separating alloys of different compositions. Experienced operators can use this method for identification of a number of ferrous alloys with reasonable accuracy. Knowledge of this identification method for ferrous alloys allows one to sort steels from superalloys.

When a piece of steel is held in contact with an abrasive grinding wheel that is rotating at sufficient speed, a stream of sparks is produced at the point of contact. The sparks are the result of abrasion of infinitesimal particles of metal by the grinding wheel, although the heating produced by this action will barely succeed in heating the projected metal particles to a red heat. These metal particles that are torn away by the grinding wheel become incandescent in the air, because oxidation is intense, initial heating is high, and the particles are exceedingly small. As the particles are projected through the air, the trajectory (called a carrier line) is easily followed (against a dark background).

Proficiency in spark testing requires practice in identifying the sparks and reproducing spark results, so that a given material will always exhibit the same spark patterns. Use of the same wheel, the same pressure, and constant lighting conditions are important factors in reducing the variables encountered in spark testing. In the descriptions of spark trails, a number of terms are used to describe parts of the trail. These are listed and shown schematically in Fig. 16.1.

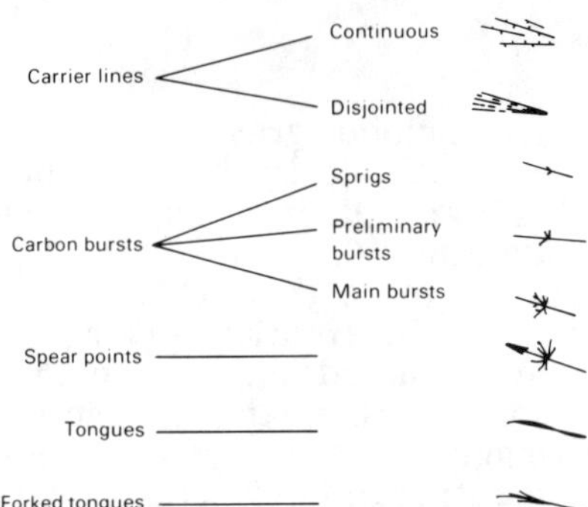

Fig. 16.1 Schematic representation of spark-testing terminology

The spark burst, or "carbon spark," is the most useful characteristic of the spark stream, because the variations in the number and intensity of the bursts indicate the changes in carbon content of the alloy. (See Fig. 16.2.)

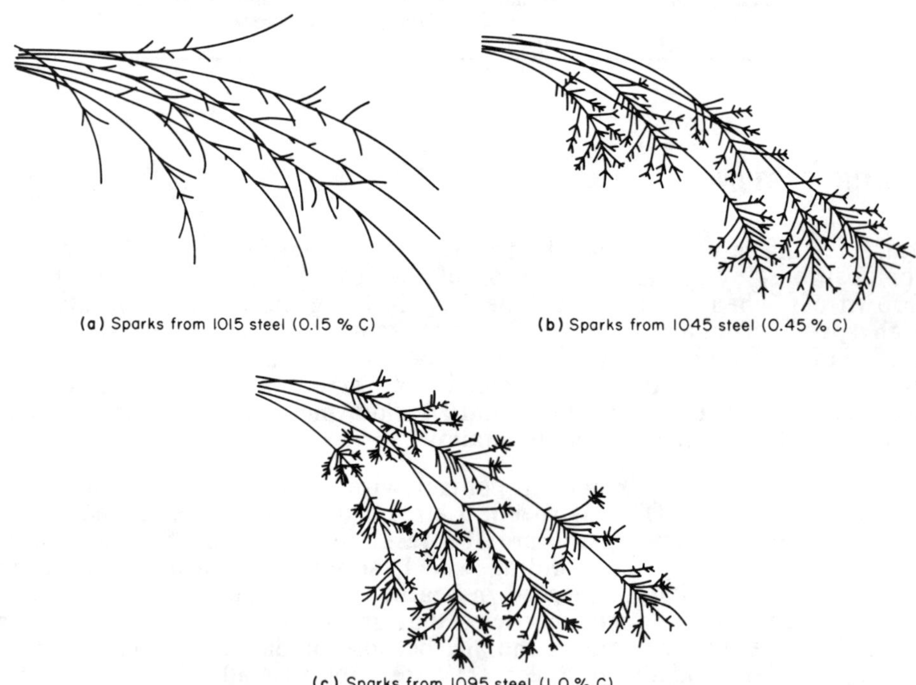

(a) Sparks from low-carbon steel, showing slight forking effect. (b) Sparks from medium-carbon steel, showing pronounced bursts. (c) Sparks from high-carbon steel, showing the intensity of bursts that is characteristic of steel having high carbon content.

Fig. 16.2 Effect of carbon content of steels on the spark pattern or stream

Table 16.4 Spark-stream characteristics of metals and alloys

Material	Description of spark system
Normal carbon steel	Heavy dense sparks 45.5 to 61 cm (18 to 24 in.) long that travel completely around the grinding wheel. Sparks are white to straw colored with main bursts throughout.
400 series chromium stainless steel	Sparks are not as heavy or dense as in normal carbon steel. Sparks are 35.6 to 45.5 cm (14 to 18 in.) long, travel completely around the grinding wheel, and are orange to straw colored, ending with a forked tongue. Preliminary bursts and few main bursts.
300 series 18-8 stainless steel	Sparks are not as heavy or as dense as those of normal carbon steel. Sparks are 30.5 to 45.5 cm (12 to 18 in.) long, travel completely around the grinding wheel, and are orange to straw colored, ending in a straight line with few, if any, bursts.
310 series 25-20 stainless steel	The spark stream is thin and from 100 to 150 mm (4 to 6 in.) long. Sparks are orange to red in color, do not travel around the grinding wheel, and there are no bursts.
Nickel and cobalt superalloys	The spark stream is thin and about 50 mm (2 in.) long. The sparks are dark red in color, do not travel around the grinding wheel, and there are no bursts.

Alloys that have the same carbon content, but different amounts of alloying elements, are not always easily identified. Most of the alloying elements, however, have some influence on the characteristics of the spark stream; they may affect the carrier lines, the bursts, or the type of characteristic sparks. Alloying elements may retard or accelerate the carbon spark, or make the carrier line lighter or darker.

Descriptions and diagrammatic representations of the sparks obtained are readily available. Of the more common alloys, those containing iron or nickel give off characteristic sparks. Alloys of cobalt, tungsten, molybdenum, and titanium also give off characteristic sparks. Abbreviated descriptions and colors of spark streams obtained during scrap testing are given in Tables 16.4 and 16.5, respectively.

In making any spark test, the requirement for known standards cannot be overemphasized. These standards must be quickly available to aid the sorter in matching knowns and unknowns.

CHEMICAL SPOT TESTING

Chemical spot testing is a relatively simple, qualitative method that can be used for rapid laboratory or field identification of metals and alloys. Information on the chemical composition of the test piece is gained by observing the color change that occurs during a chemical reaction taking place on one spot on the test piece, on filter paper, or on a spot plate. The test method is not dependent on the use of auxiliary optical magnification.

An important part of chemical spot testing is the manipulation of drops of test or reagent solutions. The success of the test method is dependent upon the nature of the reagents used, together with the advantageous use of reactive conditions, so that the desired sensitivity and selectivity can

Table 16.5 Spark-stream colors of metals and alloys

Material	Spark-test color
Nickel	Coarse red
D nickel	Coarse red
Z nickel	Coarse red
Monel	Coarse red
K Monel	Coarse red
S Monel	Coarse red
Cupronickel	Coarse red
Nickel silver	None
Inconels	Very dark red
Nimonics	Very dark red
Nichrome	Fine orange-red
330 stainless (31-15)	Coarse orange-red
310 stainless (25-20)	Fine orange-red turning white
309 stainless (25-12)	Coarse light orange turning white
300 stainless (18-8)	Light and diffused
400 stainless	Very light and diffused
Cobalt	Coarse red
Tungsten	Short yellow-white
Tungsten carbide	Short yellow-white
Molybdenum	Short yellow-white
Titanium	Brilliant white
Muntz metal (75% Ni, 6% Cu, 2% Cr, rem Fe)	Coarse red

be obtained with a minimum of physical and chemical operations. Tests ordinarily are performed by using one of the following techniques:

* Bringing together a few drops of the test solution and of the reagent on porous or nonporous supporting surfaces such as paper, glass, or porcelain
* Placing a few drops of test solution on a medium (filter paper, asbestos, or gelatin) impregnated with appropriate reagents
* Placing a drop or two of reagent on a small quantity of the solid specimen
* Subjecting a drop of reagent or a strip of reagent paper to the action of liberated gases from a drop of the test solution or from a small quantity of the solid specimen
* Adding a drop of test solution to a larger volume (0.5 to 2 mL) of reagent solution and then extracting the reaction products with organic solvents

The successful application of spot-testing procedures is enhanced by: (*a*) a knowledge of the chemical basis of the test used, so that every step of the procedure is understood and executed intelligently; (*b*) strict observance of the experimental conditions; (*c*) cleanness of the laboratory or test site and equipment; and (*d*) use of the purest reagents available. If possible, tests should be repeated to ensure reproducibility, and tests on unknown materials should be compared with tests on known materials.

In spot-testing analysis, metals almost always are detected by characteristic reactions of their mobile hydrated ions. Consequently, it must be possible to make test solutions of the material being tested. Because spot tests uniformly employ more sensitive and less disputable reactions, the presence or absence of one particular substance often can be detected with as little as one drop of the dilute test solution, even though considerable quantities of other materials are present. When testing alloy steels, follow the general procedures described below.

Testing for Iron

Place one drop of 50% nitric acid on the clean metal surface and allow to react for 1 min. Add one drop of 10% potassium thiocyanate. A blood-red color identifies the presence of iron.

Testing for Chromium

Place six drops of concentrated hydrochloric acid on the clean metal surface, or on several metal filings in a small test tube; add three drops of concentrated nitric acid and three drops of water.

Add ten drops of 10*M* sodium hydroxide and five drops of 3% hydrogen peroxide. Stir with a glass rod and allow to sit 5 min before proceeding.

Place a small ball of absorbent cotton in the solution. Using a medicine dropper, draw off clear liquid through the cotton ball by pressing the tip firmly down into the cotton. Transfer five drops of liquid (for the chromium test) and five drops (for the molybdenum test described below) to each of two depressions in a spot plate. Add two drops of diphenylcarbazide reagent and stir with a glass rod for 1 min; while stirring, add one to two drops of 3*M* sulfuric acid. A bright-red color will appear and then disappear, leaving a faint-violet color that intensifies after 1 or 2 min, which indicates the presence of chromium in the metal.

Testing for Molybdenum

To the five drops of test solution set aside from the first test for chromium described above, add a few grains of solid potassium ethyl xanthogenate, stir, and add three or four drops of 3*M* hydrochloric acid. A deep-pink color identifies the presence of molybdenum.

Testing for Manganese

Place two drops of 50% nitric acid on the surface of the metal; transfer one drop of the acid test solution to a spot-plate depression; add sodium bismuthate reagent dropwise until a brown precipitate appears. Let stand 2 or 3 min and add one drop of concentrated nitric acid. A pink color confirms the presence of manganese.

Testing for Nickel

There are three tests that commonly are used to test for the presence of nickel. In one test for nickel, the chemicals are combined into two solutions, which are prepared and used as follows:

Test Solution No. 1: Combine 100 cm^3 of concentrated nitric acid, 125 cm^3 of water, and 25 cm^3 of phosphoric acid.

Test Solution No. 2: (*a*) Dissolve 10 g of ammonium acetate crystals in 30 mL of ammonium hydroxide, (*b*) prepare solution of dimethylglyoxime, and mix solutions (*a*) and (*b*).

Cut clean, white paper toweling into 50-mm (2-in.) squares and form a slight depression in the center of each square with clean fingers. Apply one drop of solution No. 1 to the surface of the metal being tested and let stand for 1 min. Place the center of the paper square over the reaction solution for 15 s, then pick off the paper without wiping. Apply two drops of solution No. 2 to the center of the paper. A pink color signifies the presence of nickel; the greater the intensity, the higher the nickel content. The nickel content can be determined to some accuracy by conducting simultaneous tests on known and unknown specimens and comparing intensities of the pink (not brown) color. If the results are not conclusive, the solutions can be checked by conducting tests on specimens having widely different nickel contents, such as 4340 and 8620 steels. The intensity of the pink color should be markedly different.

A second test for nickel is as follows: Allow one or two drops of 50% nitric acid to react to completion on the surface of the metal to be tested. Neutralize with a slight excess of zinc oxide. Add two or three drops of saturated dimethylglyoxime. A pink color identifies nickel. The minimum nickel content detectable is 0.05%.

A third test for nickel is as follows: Place two or three drops of 50% nitric acid on the surface of the metal and allow to react to completion (about 2 min). Transfer one drop of the test reaction solution to a spot-plate depression. Add one drop of phosphoric acid and stir. Add two drops of dimethylglyoxime reagent and stir. Add one drop of concentrated ammonium hydroxide. A pink-to-red precipitate identifies the presence of nickel.

Testing for Cobalt

Place two drops each of concentrated hydrochloric acid and concentrated nitric acid on the surface of the metal. Allow 2 to 3 min to react and transfer two drops of the reaction test solution to a spot-plate depression. Add two drops of sodium fluoride solution or phosphoric acid to mask the interference of iron. Allow chemicals to react while stirring until the solution is clear. Add two drops of 10% potassium thiocyanate and two drops of acetone. An aqua-green to blue-green color identifies the presence of cobalt. The color is not stable and will disappear; however, the color will return on the addition of a few drops of acetone. A pink color following the aqua-green color confirms the presence of molybdenum also.

SUMMARY

Because of the many different situations involved in recovering and using scrap, there cannot be any hard and fast rules or fixed procedures for ascertaining identity. As discussed briefly above, scrap may be identified by object recognition (usually color), apparent density, spark patterns, chemical spot-testing, and the more time-consuming and expensive analytical procedures. Certain commercial devices also are available, including fluorescent X-ray, spectrographic analyzers, portable optical emission devices, and thermoelectric sorters.

Chapter 17

Failure Analysis and Prevention

INTRODUCTION

Failure analysis and prevention is, of course, more of interest to the applications engineer than to those involved with primary processing. However, design engineers also should be aware of the basic concepts involved in failure analysis, because selection of an inappropriate superalloy may result in the eventual failure of an application.

The general procedures, techniques, and precautions employed in the investigation and analysis of metallurgical failures are covered in extensive detail in "Failure Analysis and Prevention" in Volume 11 of the 9th Edition of *Metals Handbook.*

The examples and environmental/mechanical factors discussed below represent typical failure cases caused by a wide range of conditions. They may be used as starting points for further analyses.

CORROSION

Superalloys, as a group, exhibit excellent corrosion resistance at room temperature and at moderately elevated temperatures due to the chromium contents contained in the various iron-, nickel-, and cobalt-base alloys. Superalloys containing 20% or more chromium are suitably resistant to many aggressive environments. Generally speaking, therefore, corrosion at these low-to-moderate temperatures is not a failure condition with superalloys.

Hot Corrosion

Hot corrosion, however, may be a problem with some superalloys. (See Chapter 15 on protective coatings for a review of hot corrosion.) Hot corrosion generally refers to any corrosive behavior in which the rate of attack is accelerated significantly over and above that expected from oxidation. Difficulties from hot corrosion have been experienced in gas turbines and jet engines, particularly those operating in marine atmospheres, and in petroleum-refining and petrochemical equipment.

The two most common types of hot corrosion are vanadium corrosion and sulfidation; both originate in impurities in the petroleum-base fuel or feed stock. In both types, the impurity results in molten compounds on the metal surface. If vanadium is present in the form of vanadium pentoxide (V_2O_5), complex sodium vanadates ($n Na_2O \cdot V_2O_5$) or sodium vanadylvanadates ($n Na_2O \cdot V_2O_4 \cdot m V_2O_5$) may form. If sulfur is present, the deposit depends on the source of sulfur and operating conditions. For turbines operating on sour (high-sulfur) fuel in salt air, the result is sodium sulfate (Na_2SO_4).

In both of these types of hot corrosion, the threshold temperature for damage corresponds fairly well to the melting point of the compound. The molten compound fluxes, destroys, or disrupts the normal protective oxide. In petroleum processing, such as cracking and reforming, sulfur may be present in the feed in many forms, including organic sulfur compounds and hydrogen sulfide. The damage mechanisms in such cases are often analogous to oxidation, but are much more rapid because mass transport is much more rapid through sulfide scales, which are not very protective.

Remedial measures for hot corrosion consist of the elimination, whenever possible, of the impurities in fuel or feed, limiting operating temperatures (which is not practical in aircraft turbine engines, because the higher the operating temperatures, the higher the thrusts that the engines can produce), selection of more resistant alloys, and the use of protective coatings. Aluminum diffusion coatings have met with some success where hydrogen sulfide is the problem. In aircraft turbine applications, aluminum coatings on superalloys generally have prevented sulfidation corrosion. Relative to marine and industrial turbine engines operating with higher sulfur content fuels and often in salt environments, the clad NiCrAlY or CoCrAlY type coatings have been successful in preventing sulfidation attack. The main method of coping with vanadium corrosion is to avoid using vanadium-bearing fuel in turbines operating above the threshold temperature.

FAILURE IN ARC-WELDED SUPERALLOYS

Arc welding of superalloys has much in common with the arc welding of stainless steels. Nickel-base and cobalt-base superalloys, as well as iron-base alloys, can be welded by arc-welding processes; specific procedures vary with composition and strengthening mechanism. Joint design, edge preparation, fit-up, cleanness of the base metal and filler metal, shielding, and welding technique all affect weld quality and must be carefully controlled to prevent porosity, cracks, fissures, undercuts, incomplete fusion, and other weld imperfections. (See Chapter 13 on welding for a more complete review of arc welding.)

Fatigue Fracture

To illustrate a typical fatigue failure in superalloys, consider the fatigue fracture of a gas turbine inner combustion chamber case assembly illustrated in Fig. 17.1. The case and stiffener of an inner combustion chamber case assembly failed by completely fracturing circumferentially around the edge of a groove arc weld that joined the case and stiffener to the flange (Fig. 17.1a). The assembly consisted of a cylindrical stiffener inserted into a cylindrical case that were both welded to a flange. The case, stiffener, flange, and weld deposit were all made of the nickel-base

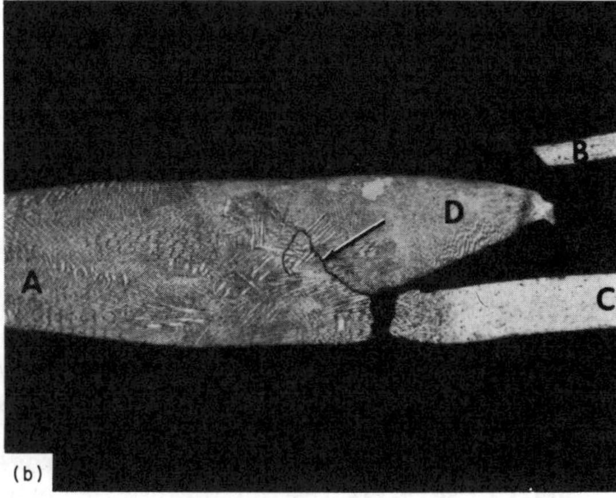

(a) Exterior surface of the assembly showing the circumferential fracture of the case (arrow). Magnification, 0.5×. (b) Section through the fracture showing the weld (region A) that originally joined the flange to the case (region B) and the stiffener (region C). The external repair weld (region D) had only partial fusion with the earlier bead. The arrow shows a film of oxide slag at the interface. Etched with 2% chromic acid plus HCl. Magnification, 10×.

Fig. 17.1 INCO 718 inner combustion chamber case assembly that fractured due to fatigue in the weld joining the flange to the case and stiffener

superalloy INCO 718. Previous to failure, a manual arc weld repair had been made along almost the entire circumference of the original weld.

Microscopic examination of the fracture site revealed unfused weld-metal surfaces and severe reductions (undercuts) in thickness of the stiffener in many areas. The thickness of the case also had been undercut in several areas. Fatigue cracks had originated at multiple sites along the weld interfaces of the case and stiffener. Also, the groove weld contained areas of mismatch that were greater than those allowed by the specifications.

Metallographic specimens from sections of the fracture site showed the double weld beads that were caused by a repair weld. Examination also indicated that many of the interfaces were unfused, as evidenced by films of oxide dross (Fig. 17.1b).

Conclusions

Failure occurred due to fatigue emanating from multiple origins caused by welding defects. The combined effects of undercutting the case wall, weld mismatch, and unfused weld interfaces contributed to high stress concentrations that generated the fatigue cracks. Ultimate failure was by tensile overload of the sections partly separated by the fatigue cracks.

Recommendation

Correct fit-up of the case, stiffener, and flange is essential, and more skillful welding techniques should be used to avoid undercutting and unfused interfaces.

HYDROGEN EMBRITTLEMENT

The term "hydrogen damage" has been used to designate a number of processes that occur in metals, by which the load-carrying capacity of the metal is reduced due to the presence of hydrogen, often in combination with residual or applied tensile stresses. Although it occurs most frequently in carbon and low-alloy steels, many metals and alloys are susceptible to hydrogen damage. Hydrogen damage in one form or another can severely restrict the use of certain materials.

High-Strength Steel

For a better understanding of the hydrogen embrittlement problem, the following discussion relative to hydrogen in steel is presented. When high-strength steel containing hydrogen is stressed in tension, even if the applied stress is less than the yield strength, it may fail prematurely in a brittle manner. This type of hydrogen damage occurs most often in high-strength steels, primarily quenched-and-tempered steels and precipitation-hardened steels. The presence of hydrogen in steel reduces the tensile ductility and causes premature failure under static load that depends on the stress and time. This phenomenon is known as hydrogen embrittlement.

Steel can be embrittled by a very small amount of hydrogen, often a few parts per million, and hydrogen may come from various sources. Unlike stress-corrosion cracking, cracks caused by hydrogen embrittlement usually do not branch, and the crack path can be either transgranular or intergranular. Failure by the hydrogen embrittlement mechanism is accompanied by very little plastic deformation, and the fracture mode is usually brittle cleavage or quasi-cleavage fracture. The susceptibility of a material to hydrogen embrittlement generally increases with increased strength level. For a given hydrogen content, the tendency to embrittlement increases with decreased strain rate, and the embrittlement is most prevalent at room temperature. The cracking tendency decreases with increasing temperature, and above 200 °C (390 °F), hydrogen embrittlement is not a problem.

Superalloys

In general, hydrogen embrittlement has not been a serious problem with superalloys, mainly because they usually operate at elevated temperatures. However, laboratory tests have shown that certain superalloys are susceptible to embrittlement when exposed to highly oxidizing environments, or to pure hydrogen at a pressure of 34 MPa (5 ksi) and a temperature of 680 °C (1250 °F). Figure 17.2 illustrates hydrogen embrittlement failure of bolts fabricated from an iron-base superalloy, Unitemp 212, which contains 16% Cr, 25% Ni, and 4% Ti.

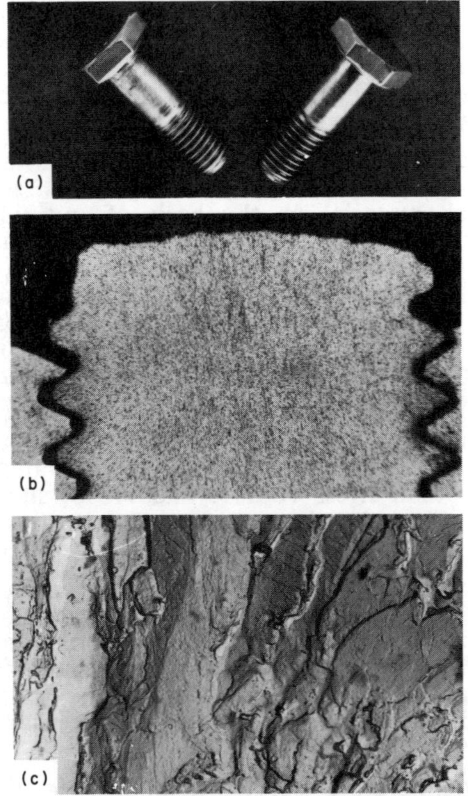

(a) Type of bolts that failed. Magnification, 0.75×. (b) Cross section through failed bolt that broke in the unthreaded shank (top) near the last full thread. Magnification, 9×. (c) Transmission electron fractograph from a plastic-carbon replica of a fracture surface of a bolt showing quasi-cleavage. Magnification, 7000×.

Fig. 17.2 Unitemp 212 bolts that failed in service from hydrogen-induced delayed cracking

HIGH-TEMPERATURE FAILURE

The greatest structural problem concerning superalloys relates to elevated temperature operating conditions. In service at elevated temperatures, the life of a metal component subjected to either static or dynamic loading is predictably limited. In contrast, at lower temperatures, and in the absence of a corrosive environment, the life of a part is unlimited under static conditions, provided the operational loads do not exceed the yield strength of the metal. Stress imposed at elevated temperature, however, produces a continuous strain in the component and results in creep. By definition, creep is time-dependent strain or deformation that occurs under stress at elevated temperature. After a period of time, creep terminates in fracture called "stress-rupture." The conditions of temperature, stress, and time under which creep and stress-rupture failure occurs depend on the alloy and the service environment. Consequently, elevated temperature failure may occur over a wide range of temperatures. In general, creep occurs at a temperature slightly above the recrystallization temperature of the metal involved, when the atoms become sufficiently mobile to allow time-dependent rearrangement of structure.

Elevated temperature behavior - the mechanical strength of the metal becoming limited by creep rather than by yield strength - must be determined for each material on the basis of individual characteristics. The primary metallurgical factor affecting stress-rupture behavior is the transition from transgranular to intergranular fracture, another important behavioral distinction between low and high temperatures. At low temperatures, the grain-boundary regions are stronger than grains, and thus, deformation and fracture are usually transgranular. At high temperatures, grain boundaries are weaker than grains, so deformation and fracture are largely intergranular.

Creep and stress-rupture failure are common high-temperature problems with superalloys. However, parts can fail at elevated temperatures for many reasons other than stress-rupture and creep, such as low-cycle and high-cycle fatigue, thermal fatigue, tension overload, and combinations of these conditions.

Another characteristic of high-temperature service is metallurgical instability. Stress, time, and temperature may change the structure of some superalloys (usually forming hard brittle compounds), which reduces strength and ductility, thus contributing to premature failure. Changes in grain-boundary structures also may contribute to premature failure. Microstructure and alloy stability are discussed in Chapters 3 and 6, respectively.

Satisfactory high-temperature service depends, to a large degree, on proper superalloy selection. Superalloys with adequate creep and stress-rupture strength - as well as good stability characteristics for the service conditions of stress, temperature, and environment - must be selected if the part is to operate satisfactorily.

Index